C.

July 23, 1981

(Corrected Copy)

Ecologic Atlas of Benthic Foraminifera of the Gulf of Mexico

Ecologic Atlas of Benthic Foraminifera of the Gulf of Mexico

C. Wylie Poag
U.S. Geological Survey
Woods Hole, Massachusetts

MARINE SCIENCE INTERNATIONAL
Woods Hole, Massachusetts

To Martha, Tracy, Marla, Graham, Mary Elizabeth,
and all my Gulf Coast colleagues

Library of Congress Catalog Card Number: 81-3720
ISBN: 0-87933-900-4

83 82 81 1 2 3 4 5
Manufactured in the United States of America.

Library of Congress Cataloging in Publication Data

Poag, C. Wylie.
Ecologic atlas of benthic foraminifera of the Gulf of Mexico.
Bibliography: p.
Includes indexes.
1. Foraminifera—Mexico, Gulf of—Ecology. 2. Foraminifera—Mexico, Gulf of—Geographical distribution. 3. Protozoa—Ecology. 4. Protozoa—Geographical distribution. 5. Protozoa—Mexico, Gulf of—Ecology. 6. Protozoa—Mexico, Gulf of—Geographical distribution. I. Title.
QL368.F6P6 593.1'2'0916364 81-3720
ISBN 0-87933-900-4 AACR2

Distributed world wide by Academic Press,
a subsidiary of Harcourt Brace Jovanovich,
Publishers.

Contents

Preface

The Gulf of Mexico has been one of the most intensely studied areas in the world for the purpose of understanding the distribution and ecology of benthic foraminifera. The northern and eastern continental shelves and their adjacent bays, lagoons, and estuaries have received most attention. But up to now, very little has been published regarding foraminiferal distribution on the southern shelf and the abyssal plains. During the past ten years, I have examined extensive collections from these poorly known areas and collected numerous additional samples. For this book, I have synthesized all of the available large-scale ecologic and distributional data in order to present an interpretation of the generic distribution patterns of the benthic foraminiferal community of the entire Gulf of Mexico. These data and their interpretation are presented in Part I.

Coincident with the collection and analysis of these new distributional data, the development of the modern scanning electron microscope has provided a means of obtaining detailed three-dimensional views of the foraminiferal test's external morphology. In Part II, this volume presents sixty-four plates of scanning electron photomicrographs and descriptions of the principal species that inhabit the Gulf of Mexico. This is not a monograph of all the species, a work that would encompass many hundreds of pages; instead it presents a selection of most of the more abundant or ecologically diagnostic species that belong to the predominant genera in the Gulf. These photomicrographs should help to stabilize the basic taxonomy of these and related species, which is essential for a meaningful interpretation of their relationships to environmental properties.

This book is designed to aid students and teachers of geology, biology, and oceanography, micropaleontologists in government and industry laboratories, and other researchers and consultants who have an interest in ecology or paleoecology. Although these specialists will benefit chiefly from the scientific attributes of the work, I hope that they, and casual readers as well, will appreciate the aesthetic qualities of the photomicrographs.

A host of friends and colleagues have contributed to this book, both indirectly and directly. First among the indirect contributors are Lyman D. Toulmin, Florida State University (retired); Harold V. Andersen, Louisiana State University (retired); Hubert C. Skinner, Tulane University; and Wilburn H. Akers, Chevron Oil Company. They kindled my interest in foraminifera, provided my formal training in micropaleontology, and always encouraged me enthusiastically in my educational and professional endeavors. Direct contributors include Richard Rezak, William M. Bryant and Arnold H. Bouma at Texas A&M University, and Alan D. Hecht of West Georgia College who provided most of the samples, and numerous colleagues at other institutions who have contributed their time and knowledge.

Those who have provided typing, editing, and critical review are too numerous to list, but their efforts are deeply appreciated. Rebecca A. Belastock, John Dunlavey, Robert F. Commeau, John E. McLane, Susan McCarthy, and Dann S. Blackwood, all of the U.S. Geological Survey, and Linda

Golder of the Marine Biological Laboratory are responsible for the final plate photography. Doris Low of the U.S. Geological Survey prepared numerous samples. The final stages of this work were completed with the encouragement of Joseph E. Hazel, who was then chief of the paleontology and stratigraphy branch of the U.S. Geological Survey. Special thanks are due to Peggy Hempenius for typing the final version (and several earlier versions), to Patricia Forrestel, Lanci Valentine, and Linda Sylwester who prepared the maps, and to my colleagues William A. Berggren, Wayne D. Bock, Thomas M. Cronin, Charles E. Pflum, and Ruth Todd for their thorough reviews.

C. Wylie Poag

Ecologic Atlas of Benthic Foraminifera of the Gulf of Mexico

Part I

Ecology and Distribution of Predominant Genera

INTRODUCTION: ENVIRONMENTAL CHARACTERISTICS OF THE GULF

The Gulf of Mexico is classified as a small ocean, the ninth largest on the globe. Its waters cover approximately 1,540,000 km^2, constitute a volume of more than 300,000,000 km^3, and reach depths of 4,376 m. In order to understand the relationships between the benthic foraminiferal communities and the varied environments of this vast region, one must first examine the major environmental properties. These properties include the physiography, general geology, and climate of the surrounding shores, as well as the major environmental features of the marine realm: bottom physiography and geology, climate, circulation, water masses, and the resulting biogeographic character. Environmental data are plentiful for the northern margin of the Gulf. This coastal region is heavily populated and includes such large cities as Tampa, New Orleans, and Houston, and the coastal plain and continental shelf constitute one of the richest domestic petroleum provinces in the United States. Consequently extensive climatological, geological, and biological studies have been carried out here by numerous agencies and institutions for many years, producing vast quantities of detailed information.

Conversely the southern margin of the Gulf has received little attention. The Mexican coast is more sparsely populated, and the petroleum activity there is still in infancy, though it is expanding rapidly. As a result, the environmental data for this southern area are sparse and are suitable only for broad generalizations. The central (deeper) parts of the Gulf also have been relatively neglected. The general geologic framework, climate, sediment distribution, and pattern of surface currents are well known, and more specific data have been acquired for scattered localities, but knowledge of its physical, chemical, and biological oceanography on a gulf-wide scale is rudimentary.

PHYSIOGRAPHY AND GEOLOGY OF THE GULF COASTAL PLAIN

A brief review of the physiography and geology of the adjacent coastal plain is important because many of the sediments and dissolved minerals that provide substrates and test materials for the foraminiferal assemblages are derived from the coastal rocks. Furthermore, the courses of the numerous rivers that freshen the coastal waters and form extensive estuarine and deltaic biotopes are greatly influenced by the topographic and geologic features of the coastal plain. Many of the coastal physiographic features also extend onto the continental shelf, where they influence foraminiferal distributions. Murray (1961) has provided a detailed review of the physiography and geology of the coastal plain.

The Gulf of Mexico lies within a semienclosed basin (Figure 1). The eastern margin of the Gulf is formed by the Florida peninsula, a region of coastal lowlands and plains,

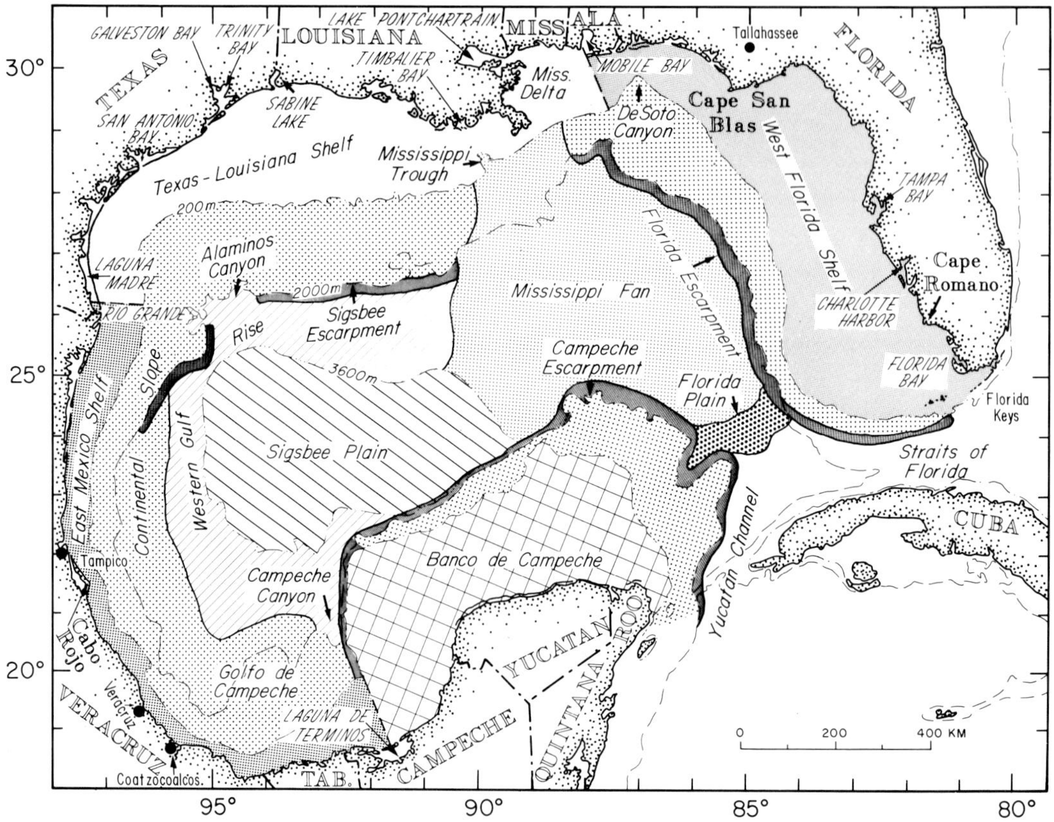

Figure 1
Physiographic features and provinces of the Gulf of Mexico (modified from Bouma, 1968; Martin and Bouma, 1979).

approximately 225 km across. Thick beds of Tertiary limestone underlie most of the peninsula and are generally covered by quartz sands of Pleistocene origin. The surface of this region is drained on the Gulf side principally by the Caloosahatchee and Suwannee rivers (Figure 7), but underground cavities in the soluble limestones provide additional drainage (Kohout et al., 1977).

The Gulf shoreline of Florida is complex. It begins at the south with the Florida Keys, a crescentic chain of low-relief carbonate islands, which form the boundary between shallow Florida Bay and the deep Straits of Florida. Toward the north, a series of vast island-dotted mangrove and cypress swamps lines the shore up to about Charlotte Harbor. Charlotte Harbor and Tampa Bay are the two major west Florida estuaries, formed by the Peace and Hillsborough rivers, respectively, and each bay is rimmed by sandy quartzose barrier beaches. Narrow quartz sand beaches compose the bulk of the remaining west Florida shore.

A much wider coastal plain extends from the panhandle of Florida to Tampico, Tamaulipas. That part adjacent to the Gulf is composed of deltaic and coastwise plains and marshes. Inland is a series of coastwise alluvial terraces of Pleistocene age. The innermost edge of the coastal plain contains belted topography developed on the outcropping bands of Tertiary and Upper Cretaceous sedimentary strata. At its widest point, the coastal plain extends inland for more than 800 km, reaching southern Missouri and Kentucky. This segment, therefore, provides the greatest variety and volume of sedimentary particles to the Gulf basin (van Andel, 1960).

The principal drainage of this wider segment of coastal plain begins to the east with the Flint-Chattahoochee-Apalachicola river system of northern Florida and the Alabama-Tombigbee-Mobile river system of Alabama (Murray, 1961). These rivers arise in the Appalachian uplands of Alabama and Georgia. Through the middle of the coastal plain courses the Mississippi River, which enters the Gulf through a huge birdfoot delta system that forms the southeasternmost extension of Louisiana. This is by far the most significant river system entering the Gulf, deriving its waters from the entire heartland of the United Stages between the Appalachian and Rocky Mountain chains (41 percent of the total area of the contiguous United States is drained; >5 million km^2). The annual contribution of Mississippi River sediments to the Gulf of Mexico amounts to more than 600 million metric tons (Moody, 1967). Westward from the Mississippi Delta, the Sabine, Neches, Trinity, Brazos, Colorado, Guadalupe, San Antonio, Rio Grande, and several smaller rivers wind across the broad Texas plain, carrying Cenozoic and Cretaceous sediments to the western Gulf.

The northern shoreline of this wider segment of coastal plain begins in the panhandle of Florida as a series of narrow sandy quartzose beaches, dunes, barrier islands, and low coastal marshes. These give way westward to the large Mobile Bay estuary and the coastal bays and marshes surrounding the Mississippi Delta. The Texas rivers form a series of marshy estuarine embayments and lagoons behind extensive sandy quartzose barrier beaches. These estuaries entrap much of the riverborne detritus, permitting only a small volume to reach the continental shelf (van Andel, 1960).

The Mexican coastal plain from about Tampico to below Laguna de Terminos is much narrower than the other coastal segments. The maximum width of this segment is 150 km. The inner edge of the coastal plain here is bordered by the Sierra Madre Oriental and by the Chiapas Highlands. Tertiary sedimentary strata underlie the bulk of the Mexican coastal plain, but at its inner edge and in the adjacent mountains, Cretaceous and Jurassic sedimentary beds crop out. At two locations, one just north of Tampico and the other southeast of Veracruz, igneous and/or volcanic rocks reach the shore of the Gulf and form a rocky coast. The principal rivers, the Pánuco and Tamesí, have their headwaters in the Sierra Madre Oriental and bring relatively unaltered Mesozoic sediments into the western Gulf. Southwest of Laguna de Terminos, the Rios Grijalva and Usumacinta have constructed a vast alluvial plain similar to that surrounding the Mississippi Delta.

From the Rio Grande to Laguna de Terminos, the Mexican shoreline is largely a series of narrow quartz sand beaches and dunes interrupted by elongate shallow lagoons (lagunas), barrier bars, and river-mouth marshes. The major exception is the stretch of volcanic rocks between Veracruz (Alvarado) and Coatzocoalcos.

East of Laguna de Terminos, the broad limestone lowlands of the Yucatán peninsula constitute the southeastern rim of the Gulf. This area is largely drained underground through systems of caverns, sinkholes, and other carbonate solution features. No significant rivers are present in this vast region, but a few small streams and lakes occur.

The western shore of the Yucatán peninsula is rather uniform, consisting of carbonate sand beaches interrupted by minor irregularities and narrow marshes. The northern coast, however, is embayed by an extensive laguna, which is protected by barriers of carbonate sand. Prominent emergent reefs dot the shelf edge off Yucatán.

PHYSIOGRAPHY OF THE GULF FLOOR

The physiography of the Gulf floor largely determines the distribution routes and final location of siliciclastic deposits debauched by the coastal rivers. Submerged physiographic features also greatly influence the distribution and stability of sublittoral benthic communities, because the width, depth, and topographic relief of the continental shelves modify the effects of currents, waves, tides, and light penetration. The large physiographic features of the Gulf are well known (Figure 1; Ewing, Ericson, and Heezen, 1958; Bouma, 1968; Bryant et al., 1968; Bryant et al., 1969; Ballard and Uchupi, 1970; Bergantino, 1971; Garrison and Martin, 1973; Stuart and Caughey, 1977; Martin and Bouma, 1979). Seismic and fathometer profiles of the bottom have been collected from every part of the open Gulf, and many areas have been studied in great detail. For purposes of this synthesis, I describe chiefly those physiographic features that appear to have some influence on the distribution of benthic foraminifera. For a more detailed description of the submerged physiography of the Gulf, refer to the recent summary by Martin and Bouma (1979).

Seaward of the series of bays, estuaries, barrier island lagoons, mangrove swamps, and river-mouth deltas, three broad, gently sloping segments of the continental shelf extend out to about the 200 m depth (locally the shelf edge is as shallow as 60 m; Figure 1). Off Florida is the West Florida Shelf; the Texas-Louisiana Shelf extends from Mobile Bay to the Rio Grande; and the Campeche Shelf (Banco de Campeche) borders the northern and western parts of the Yucatán peninsula. The generally smooth surfaces of these continental shelves are interrupted by submerged or emergent banks and reefs and by submerged ancient shoreline features (Ballard and Uchupi, 1970). Between the broad Banco de Campeche and the Texas-Louisiana Shelf is the narrow, more steeply sloping East Mexico Shelf, which is separated by only a narrow coastal plain from the Sierra Madre Oriental.

Between depths of approximately 200 m and 2000 m, the more steeply sloping continental slope rims the Gulf (Figure 1). The continental slope off Louisiana, Texas, and Campeche is quite rugged; numerous submerged banks, ridges, domes, and basins have been formed largely by intruding subsurface masses of salt. On the continental slope of eastern Mexico, a series of elongate, parallel ridges have been formed by compressional folding of thick sedimentary beds that lie above an undeformed salt layer (Buffler et al., 1979; Watkins et al., 1979). Four major submarine canyons indent the outer margin of the continental slope: the DeSoto Canyon off northern Florida; the Mississippi Trough off central Louisiana; the Alaminos Canyon off Texas; and the Campeche Canyon along the western margin of the Campeche Shelf. The basinward edge of the continental slope everywhere except off the Mississippi Delta and eastern Mexico is formed by steep escarpments: the West Florida Escarpment, the Sigsbee Escarpment, and the Campeche Escarpment.

Next, sloping more gently into the center of the Gulf basin, is the continental rise. It is rather uniformly smooth on the west and northwest (Western Gulf Rise), but eastward a huge pile of Mississippi River sediments has formed the cone-shaped Mississippi Fan. The deepest part of the basin (>3,600 m) occupies the west-central part of the Gulf. This relatively flat region is called the Sigsbee Plain. Even here, however, the mobile underlying salt beds have produced the well-known Sigsbee Knolls and other bottom prominences. A relatively small area of abyssal seafloor in the southeastern Gulf, between the Florida and Campeche escarpments, is known as the Florida Plain. Several sediment-capped knolls are present here also, but they appear to be of igneous origin (Pyle et al., 1969; Martin and Bouma, 1979).

At its southeastern corner, the margin of the Gulf basin is formed by a submerged ridge connecting Yucatán and the island of Cuba (Figure 1). This ridge forms a sill approximately 2000 m below sea level, which prevents the deeper waters of the Caribbean Sea from directly entering the Gulf. Between Florida and Cuba, the submerged rim of the Gulf is even shallower, rising to about 1000 m in the Straits of

Florida. This sill allows only the shallow water masses of the Gulf to flow into the Atlantic Ocean.

CLIMATE

The general climate of the Gulf of Mexico is significant because it controls the surface temperature, relative humidity, and precipitation and thereby serves as a basis for differentiating biogeographic provinces. Of these factors, temperature is thought by many to have the strongest influence on the distribution of marine organisms because it directly affects the rate of chemical reactions in the water and within the organisms, exerting powerful control over activity rates, metabolic functions, reproduction, and growth rates.

The Gulf of Mexico lies between latitudes 18° and 30° N, a tropical to subtropical region of generally abundant rainfall and high humidity (Figure 2; Trewartha, 1954). Five rather clearly defined regions of different continental climate can be recognized in the surrounding coastal plain. The largest single climatic region begins with the Florida peninsula north of Charlotte Harbor, and extends through Texas to just above its border with Tamaulipas. It is a humid subtropical region with average summer temperature of around 25° to 27°C. Normal diurnal temperature variation is 8° to 17°C. Throughout most of this region, rainfall is the dominant form of precipitation and averages 100 to

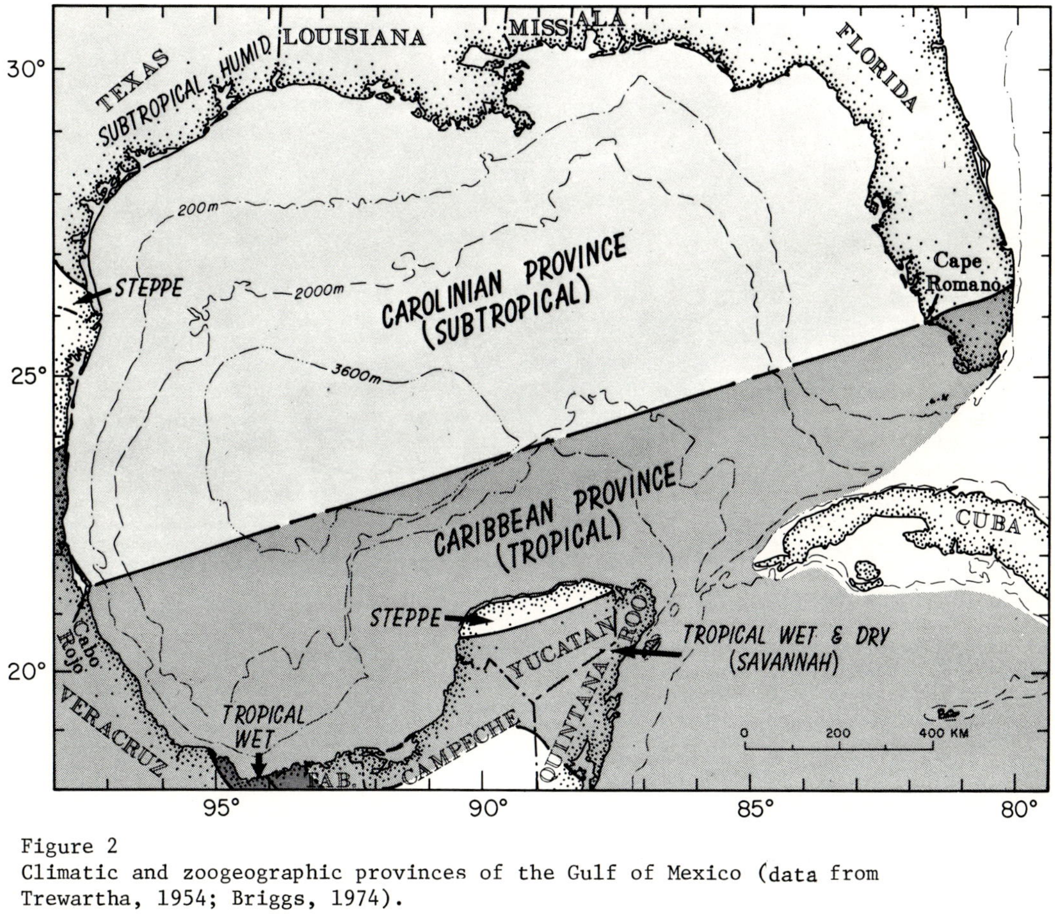

Figure 2
Climatic and zoogeographic provinces of the Gulf of Mexico (data from Trewartha, 1954; Briggs, 1974).

150 cm per year, except in two places. The first, near the coast, from the Mississippi Delta to about Tallahassee, Florida, is a zone of higher rainfall that averages 150 to 200 cm per year. The second place is the southwestern Texas coast, a much drier area, averaging only 50 to 100 cm per year.

From south of the Rio Grande to just north of Tampico (and including a part of the northwestern coast of the Yucatán peninsula), an even drier region is classified as a low-latitude steppe. The climate here is transitional between the low-latitude inland desert and the more humid coastal area to the south. Rainfall and temperatures are, accordingly, highly variable.

The rest of eastern Mexico, the bulk of the Yucatán peninsula, and southern Florida have a tropical wet and dry (savanna) climate. Diurnal temperatures normally vary by about 3° to 8°C. During the dry season (November to May), temperatures may range from 16° to 40°C; during the wet season (June to October), temperatures of 21° to 35°C are normal. Precipitation, most of which falls during the summer, ranges from 75 cm to more than 150 cm per year.

A narrow strip of the Veracruz-Tabasco coast just west of Laguna de Terminos has a tropical wet (rain forest) climate. The temperature is uniformly high (25° to 27°C) and has little seasonal variation. Rainfall is heavy and uniformly distributed throughout the year ($\sim$250 cm per year or more).

The basic dichotomy between the northern and southern Gulf coastal climates extends also over the open Gulf (Figure 3; Greiner, 1974). In the northern Gulf (divided by a line from Tampico, Tamaulipas, to Cape Romano, Florida), surface water temperatures may vary annually by more than 11°C, but the variation generally is less than 7°C in the southern Gulf. From January to April, the northern waters average around 17° to 23°C, but in the south, the variation is from about 22° to 26°C. In contrast, the summer and fall temperatures (July to October) are nearly uniformly high (around 27° to 29°C) over the entire Gulf.

CIRCULATION AND WATER MASSES

The climatic temperature effects are measurable throughout the surface water mass, whose contents are uniformly mixed to a depth of about 100 m (Figure 4; Nowlin, 1971). This means that where the mixed layer lies over the continental shelves, temperature at any point on the sea floor is equivalent to that immediately above it at the surface. However, there is a horizontal temperature gradient (Figure 5) along which temperature increases from north to south across the northern, eastern, and western shelves. But on the southern shelf, where annual temperature variations at the surface are small, the cross-shelf temperature gradient is also small.

Second to temperature, salinity is perhaps the most influential environmental factor affecting the distribution of benthic marine organisms. Salinity exerts control not only by changing concentrations of dissolved ions but by affecting water density, the boiling and freezing points, vapor pressure, and, most important, osmotic pressure.

The salinity of coastal waters is diminished along the northern shores of the Gulf by the influx of fresh river water (see Bouma, 1976). Southward, the influence of river runoff is reduced and evaporation is greater, which results in higher salinities over the inner shelf.

Farther out on the continental shelf, the salinity pattern of the surface mixed layer is similar to that of temperature; there is no vertical gradient, but a gradient from north to south is generally present (Figure 3; Greiner, 1974). In the winter and spring, mean salinity is lowest in the north around the Mississippi Delta (32 to 33 o/oo). The low salinity isopleths swing westward across the Texas-Louisiana Shelf, reflecting the prevailing coastal current direction. Water having salinity of 36 o/oo or less is found in the northern Gulf between the mouth of the Rio Grande and the tip of Florida. Southward, the salinity increases. From January

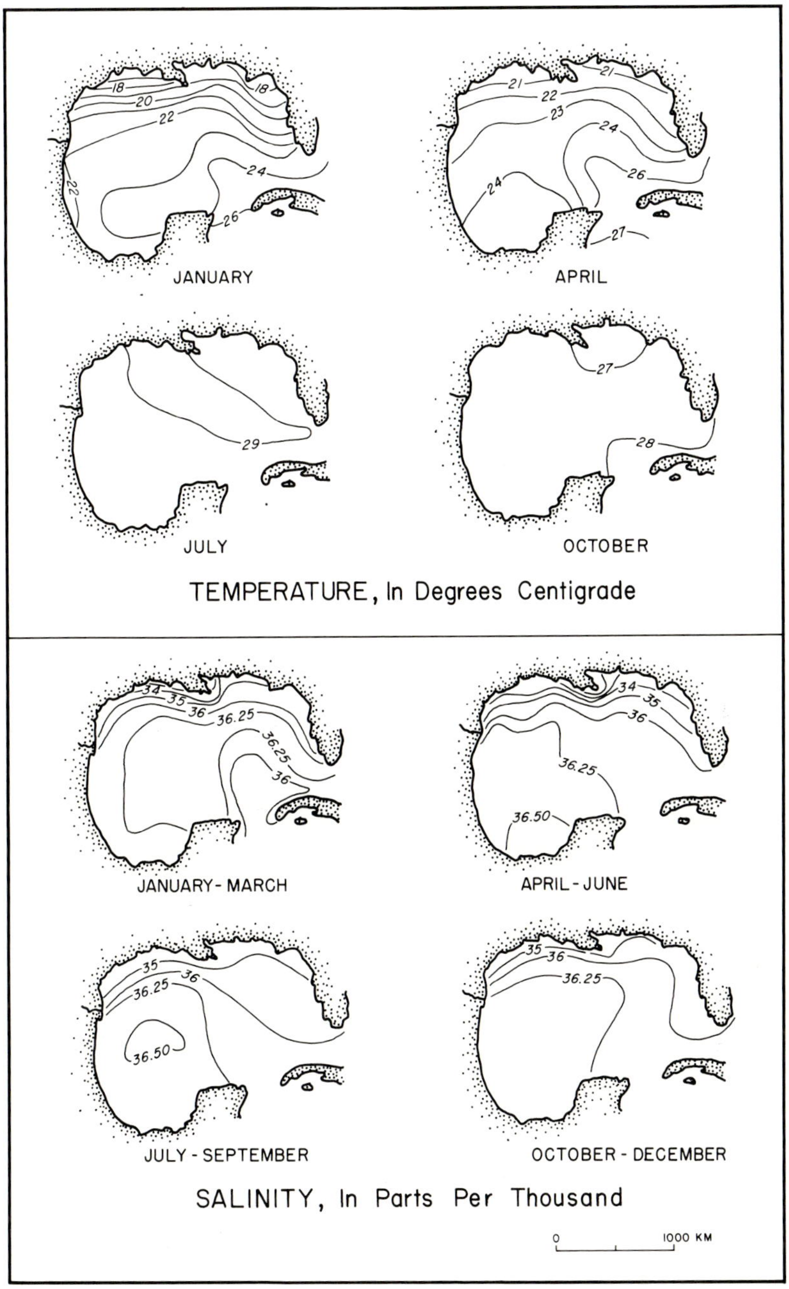

Figure 3
Distribution of mean temperature and salinity of surface waters of the Gulf of Mexico (data from Greiner, 1974).

through March, the mean salinity is highest in the central Gulf (36.25 o/oo), but from April through June, the Golfo de Campeche is the most saline area (36.5^{+} o/oo). In the summer and fall, mean surface salinity varies only one or two parts per thousand across the open Gulf (35 to 36.5 o/oo). The southwestern Gulf maintains the highest salinity, which

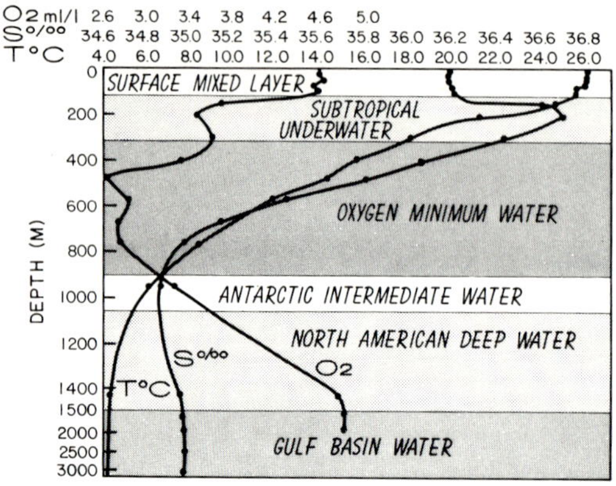

Figure 4
General stratification of water masses in the Gulf of Mexico (data from Nowlin, 1971).

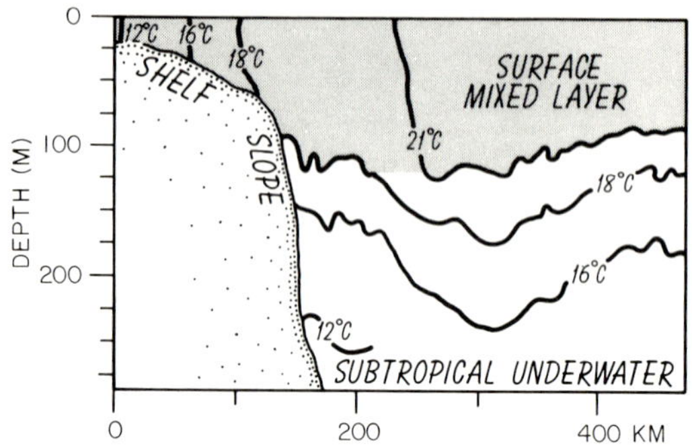

Figure 5
Temperature profile across a typical part of the northern shelf of the Gulf of Mexico (winter traverse off Galveston, Texas; data from Phleger and Parker, 1951).

is nearly constant from July through December (∿36.25 o/oo).

The major surface current in the Gulf of Mexico is the Yucatán Current, which enters from the Caribbean Sea (Figure 6; Nowlin, 1971). This warm, saline current flows across the Gulf to around lat. 25° to 28°N, and then loops southward again (the Loop Current), flowing through the Straits of Florida into the Atlantic (as the Florida Current). A prominent counterclockwise gyre arises from the Loop Current and influences the West Florida Shelf.

Most authors have shown surface currents in the northwestern Gulf flowing dominantly westward. Upon reaching the northern shelf, the currents flow westward and southward parallel to the shoreline to a nodal point off Brownsville, Texas. Here, in the summer, they impinge upon a current flowing northward off eastern Mexico. But in winter, there is a reversal of the current coming from

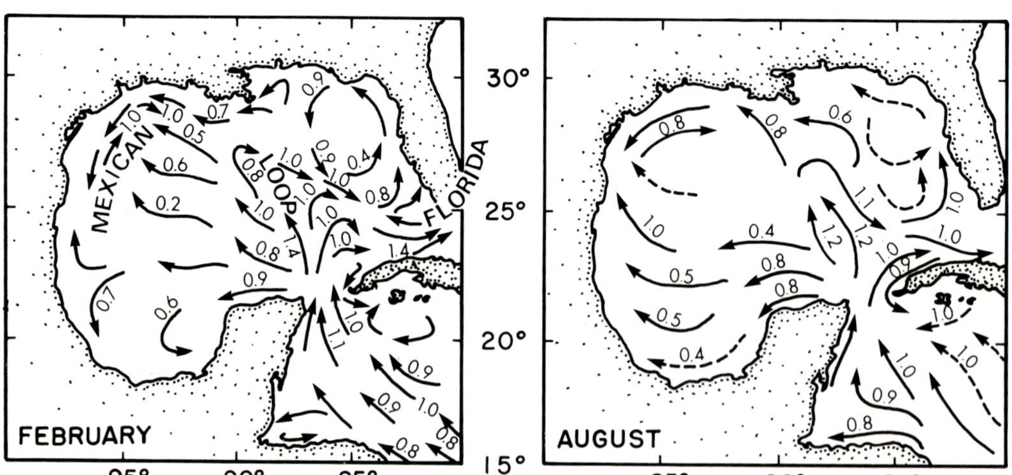

Figure 6
General distribution of major surface currents in the Gulf of Mexico (data from Nowlin, 1971). Numbers beside arrows indicate average current speed in knots.

the southwest, and it turns back eastward into the Golfo de Campeche (Nowlin, 1971). Recently Sturges and Blaha (1975) have reinterpreted hydrologic data in the western Gulf and inferred the presence of a clockwise western boundary current flowing beyond the shelf edge, which they named the Mexican Current.

Below the mixed surface water mass (at depths greater than about 100 m), the current patterns are poorly known, although several investigators have presented evidence for deep current flow. For example, Pequegnat (1972) and Pequegnat et al. (1972) presented photographic, current meter, and biological evidence of bottom currents on the distal Mississippi Fan ("East Gulf Deep Bottom Current"; see also Rowe and Menzel, 1971). Bouma (1973) later rephotographed Pequegnat's (1972) stations, found no ripple marks, and concluded that the "East Gulf Deep Bottom Current" is intermittent, touching bottom only occasionally. Moore (1973) measured a southwestward flowing countercurrent along the upper part of the continental slope off Texas and Louisiana beneath the Mexican Current.

The sequence of vertical stratification below the surface water mass has been summarized by Nowlin (1971; Figure 4). Five major subsurface water masses can be identified by their salinity, temperature, and oxygen content. Between 100 and 300 m, a core of Subtropical Underwater has temperatures around 18° to 25°C, salinity of 36.2 to 36.6 o/oo, and oxygen content around 3.8 to 3.4 ml/1. The Subtropical Underwater impinges on the bottom along the outer shelf and upper slope. Below this layer, oxygen content drops to 2.6 to 2.8 ml/1 in the Oxygen Minimum Water (300 to 900 m); salinity drops from 36.4 to 34.9 o/oo, and temperature decreases from 18° to 6°C at its base. This layer touches the bottom along the upper part of the slope. The curves for salinity and temperature converge between 900 and 1050 m in a layer called the Antarctic Intermediate Water (temperature, 5° to 6°C; salinity, 34.8 o/oo; oxygen content, 3.0 to 3.6 ml/1), which bathes the floor along the middle part of the slope. From 1050 to 1500 m, is North American Deep Water in which temperature slowly decreases to 4°C, salinity increases slightly to nearly 35 o/oo, and oxygen increases dramatically through a range of 3.6 to 4.8 ml/1 as depth increases. The base of the North American Deep Water marks the permanent thermocline. This water mass intersects the bottom along the lower slope. Below the permanent thermocline is the deepest water mass in the Gulf, known as the Gulf Basin Water. The upper 500 m portion of this water mass is directly connected with the waters of the Caribbean Sea, but the deeper parts are isolated by the 2000 m sill depth across the Yucatán Channel. The Gulf Basin Water covers the lowest part of the slope, the rise, the Sigsbee and Florida plains, and most of the Mississippi Fan with water of nearly constant salinity ($\sim$35 o/oo), temperature (4°C), and oxygen content (4.8 ml/1) and constitutes the only indigenous water mass in the Gulf. Carder et al. (1977) have shown that the Gulf Basin Water is derived in part from Yucatán Basin Water that enters the Gulf of Mexico through the Yucatán Channel. The Gulf Basin Water may also contain a component that enters the Gulf from the Atlantic through the Straits of Florida as a southward-flowing deep countercurrent (Hurley and Fink, 1963; Neumann and Ball, 1970).

SEDIMENTS

Sediments influence the distribution of marine benthic organisms chiefly through the types of substrates they provide for the organisms or their hosts, through the amounts and types of food the sediments contain, and (for foraminifera) through the types of materials the sediments might provide for constructing tests. Generally the critical aspects of sediments are the relative amounts of clay, silt, and sand; the mobility of the substrate; the rate of accumulation; and the amount (and perhaps the composition) of organic detritus.

The distribution of sediments in the Gulf of Mexico is intimately related to the coastal and submarine physiography, climate, and currents, as well as to the composition of the surrounding coastal rocks (see Lynch, 1954). The primary source of siliciclastic sediments is the Mississippi River, which has built a huge delta that nearly reaches the edge of the Texas-Louisiana Shelf (Figure 7; Gould and Stewart, 1955; Davies and Moore, 1970). Most of the Mississippi sediment load, rich in silt, clay, and organic detritus, is deposited in deep water to form the Mississippi Fan, but it also supplies the Sigsbee and Florida plains with fine silts and clays. Another component of the Mississippi sediment load is swept westward along the eastern half of the Texas-Louisiana Shelf and forms a muddy substrate (see Davies, 1972). The numerous rivers of the central Texas coast provide a mineralogically distinctive suite of muddy siliciclastic sediments, which covers most of the rest of the Texas-Louisiana Shelf and spills over onto the adjacent slope and rise.

The Rio Grande also supplies a mineralogically distinctive suite of siliciclastic sediments that forms a narrow band of muddy deposits directly basinward of its mouth. These sediments reach the northwestern corner of the Sigsbee Plain. The extensive Mexican siliciclastic sedimentary regime is supplied by numerous streams that drain the adjacent mountainous coastal region. These sandy terrigenous sediments cover the shelf, slope, and rise but

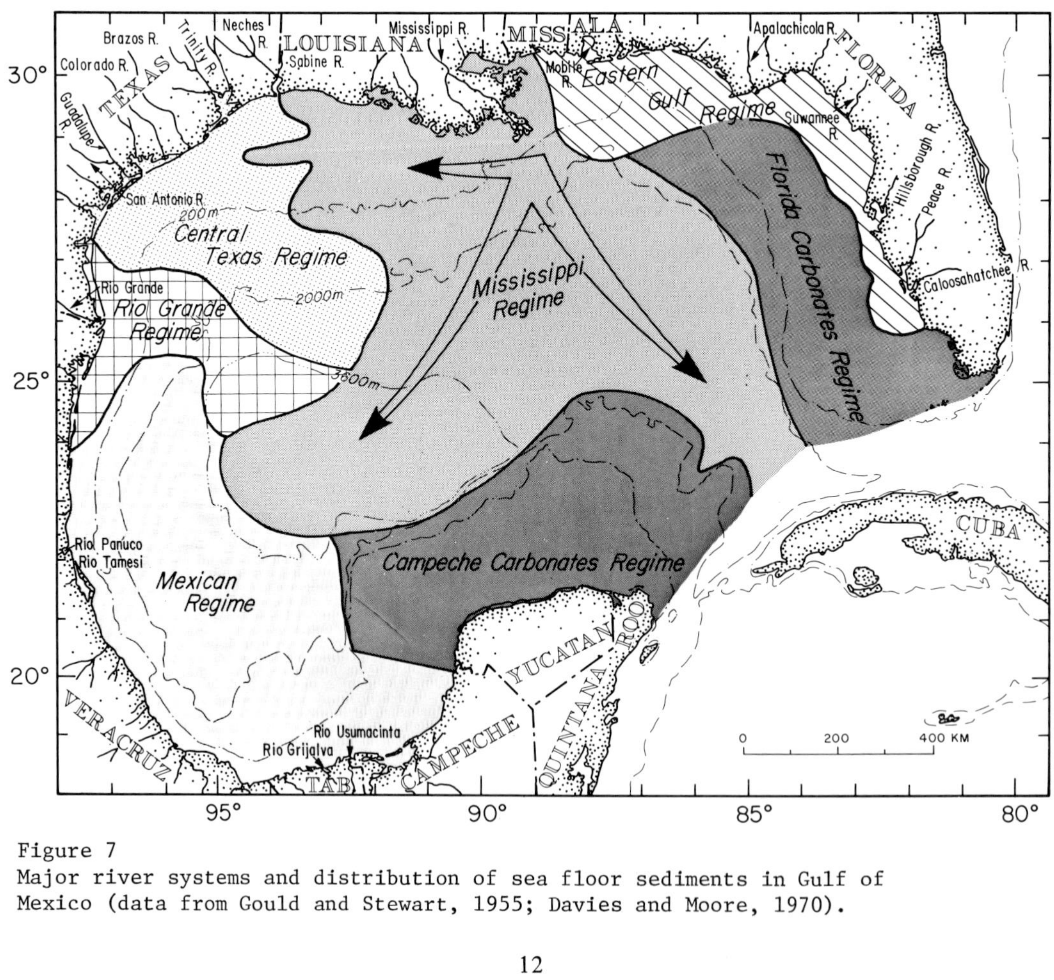

Figure 7
Major river systems and distribution of sea floor sediments in Gulf of Mexico (data from Gould and Stewart, 1955; Davies and Moore, 1970).

are not strongly manifest on the Sigsbee Plain.

The remaining siliciclastic sedimentary regime borders the eastern Gulf shore off Florida. A few rivers drain the sandy Pleistocene and upper Tertiary coastal plain rocks as far south as Charlotte Harbor but provide only small quantities of clay and silt. The substrate of the inner shelf here, therefore, is chiefly quartz sand.

Organic detritus in sediments of the western Gulf is sparse in nearshore areas (∿0.5 percent of sediment; Trask, 1953), reaches a maximum on the upper slope (∿2 percent), and then decreases to ∿1 percent on the Sigsbee Plain; clays are generally richer in organic content than sands. On the shelf, much (∿50 percent) of the organic matter in these terrigenous sediments seems to be derived chiefly from land plants (Hedges and Parker, 1976), but the organic constituents of the Sigsbee Plain deposits are almost entirely of marine origin. The Mississippi Fan appears to contain higher percentages of terrigenous organic debris than other bathyal and abyssal regions of the Gulf (Huang and Goodell, 1970; Koons and Perry, 1976).

Two distinctive areas of chiefly biogenic carbonate deposits are present in the Gulf of Mexico. The first occupies the middle and outer parts of the West Florida Shelf and the upper part of the adjacent slope. This carbonate-rich substrate abuts the Florida shoreline along Florida Bay but is separated from the shore northward by the eastern Gulf siliciclastic suite (Figure 7; Gould and Stewart, 1955). The other extensive carbonate area occupies the Banco de Campeche and Campeche Slope. This is an area of extensive production of modern biogenic carbonate deposits, especially among the living reefs of Alacrán, Arcas, Triangulos, and Arenas (Logan et al., 1969). Other more localized biogenic carbonate deposits are found around emergent reefs off Veracruz and Tampico, and on numerous submerged banks off Louisiana, Texas, Alabama, and Florida, but most of the submerged banks contain principally Pleistocene carbonate deposits (Poag, 1972).

MARINE BIOGEOGRAPHY

The shelf biotas of the Gulf of Mexico have been divided into several different biogeographic provinces and subprovinces whose number and boundaries vary according to the group of organisms considered. Pulley (1952), for example, proposed five different provinces on the basis of the distribution of bivalves. Most authors agree, though, that for the biota in general, two major provinces can be recognized. A northern province, with southern boundaries at Cabo Rojo, Veracruz, and Cape Romano, Florida, contains about 10 percent endemic invertebrates and fishes, many of which are warm-temperate or nontropical species (Figures 1, 2; Briggs, 1974). The southern boundaries of this province are nearly coincident with those bounding the two humid subtropical climatic provinces. Although the eastern and western parts of the northern biogeographic province are generally similar, some authors have pointed out differences. For example, the fish fauna is richer in the northeastern Gulf. Further, the coral-sponge communities of the sandy West Florida Shelf are replaced on the muddy Texas-Louisiana Shelf by the shrimp ground community, except on the tops of submerged carbonate banks (Hedgpeth, 1953; Bright and Pequegnat, 1974). The northern Gulf biogeographic province is considered to be a western extension of the Carolinian Province, the eastern part of which extends along the U.S. Atlantic Continental Shelf from Cape Canaveral to Cape Hatteras (Briggs, 1974). Of the two Gulf provinces, the northern one is richer in species of fish and invertebrates and has more endemic forms. With respect to invertebrate faunas, the Gulf seems to have served as the evolutionary center for shelf-dwelling species of the Carolinian Province, filtering migrants to the East Coast through the Straits of Florida during glacial intervals of the Pleistocene (Briggs, 1974).

The southern biogeographic province of the Gulf of Mexico belongs to the tropical Caribbean Province and is typified by emergent (or shallow submerged) reefs of living scleractinian corals, calcareous algae, and their associated biota. Similar reef associations are present on some of the carbonate banks near the shelf edge off Texas and Louisiana (Poag and Tresslar, 1981). Only the deepest water reef organisms are present on the northern banks, however, because the shallowest bank supporting an extensive living reef (West Flower Garden Bank) is 20 m below sea level (Bright and Pequegnat, 1974).

PREVIOUS BENTHIC FORAMINIFERAL STUDIES IN THE GULF

Following the preliminary investigations by Goës (1896), Flint (1899), Cushman (1922), and Kornfeld (1931), critical study of the distribution and ecology of benthic foraminifera from the Gulf of Mexico was accelerated by Lowman (1949). The 1950s saw a burst of investigations supported by the Gulf Coast petroleum companies, and most of the broad-scale studies stem from this period. Every foraminiferologist is familiar with the classical works of Phleger and Parker (1951), Parker (1954), and Bandy (1954, 1956), which demonstrated that extensive coastwise bands of distinct foraminiferal associations occur along the Texas-Louisiana and West Florida shelves. To this data base were added Phleger's regional study of benthic foraminiferal faunas in the northwestern Gulf (Phleger, 1960a), Walton's synthesis for the northeastern Gulf (Walton, 1964a), Bock's synthesis for the southern part of the West Florida Shelf (Bock et al., 1971) and for the Mississippi-Alabama-Florida (MAFLA) Shelf (Bock, 1976), and Pflum and Frerichs's (1976) and Dignes's (1978) studies of deep-water benthic foraminifera.

Papers of more limited geographic scope include several concerning the foraminifera from submerged carbonate banks on the Texas-Louisiana Shelf: Kellough (1956), Ludwick and Walton (1957), Loep (1965), Poag and Sweet (1972), Poag (1972), Poag and Tresslar (1979, 1981). Phleger (1954, 1955) and Lankford (1959) studied the assemblages of the Mississippi Delta and vicinity. The Texas and Louisiana bays, lagoons, and estuaries received attention from Post (1951), Parker, Phleger and Peirson (1953), Phleger (1956, 1960b), Phleger and Lankford (1957), Warren (1957), Kane (1962, 1967), Waldron (1963), Wantland (1969), Poag (1976, 1978), and Otvos (1978). Mobile Bay and environs were studied by Walton (1964a), Upshaw et al. (1966) and Lamb (1972). Walton (1964b) provided data from Tampa and Sarasota bays. Bandy (1956) studied Charlotte Harbor and Tampa Bay. Moore (1957), Benda and Puri (1962), and Lynts (1962, 1965) studied the southern part of the West Florida Shelf and Florida Bay. From the southern margin of the Gulf, Davis (1964) provided a brief description of Alacrán reef assemblages, and Ayala-Castañares (1963), Segura (1963), and Phleger and Lankford (1978) studied the composition of assemblages in some Mexican lagunas. Lidz and Lidz (1966) contributed a study of the Veracruz reef foraminifera.

Miscellaneous other published studies include those of Shifflett (1961), Howard (1965), Buzas (1967), Mello and Buzas (1968), Nelson and Bray (1970), LeRoy and Hodgkinson (1975), and Hueni et al. (1978). Several unpublished theses also have treated the distribution of benthic foraminifera in the Gulf (Bock, 1961, 1967; Tresslar, 1974; Hackathorn, 1975; Anepohl, 1976; Huse, 1976).

The synthesis presented here is based upon collections from more than fifteen hundred stations from the lagoons, bays, and estuaries surrounding the Gulf and more than three thousand from the shelf, slope, rise, and abyssal plains. Much of the census data have been previously published, but four hundred new stations from the abyssal plains, Mississippi Fan, and the Mexican Slope and shelves provide additional information that is published here for the first time. Most of these new analyses come from grab and core samples taken by Brian Logan and his colleagues at Texas A&M University (see Logan et al., 1969). The rest are from cores taken by other scientists at Texas A&M. In their original form, the previously published raw data were not always compatible from author to author, particularly because of differences in morphologic definition of certain genera. I have updated the generic taxonomy used by each author in order to eliminate this incompatibility. I have omitted Phleger's (1954) data from Mississippi Sound because they are equivocal (see Phleger, 1954: 647). I also disregarded those deep-sea core-top samples that contained cool-water Pleistocene planktic species and thus obviously did not represent the modern benthic assemblage.

GENERIC PREDOMINANCE FACIES OF MODERN GULF BENTHIC FORAMINIFERA

Phleger (1960c) summarized the ecology and distribution of modern foraminifera largely on the basis of studies in the northern Gulf of Mexico. He pointed out that distinctive species assemblages could be found in the following geographic realms: marine marshes, estuaries, coastal lagoons, beaches and nearshore areas, the continental shelf, and the continental slope-abyssal plain. Subassemblages were recognized in all these areas on the basis of presence or absence of certain species and on the relative abundance of other species. Especially noteworthy was clear documentation of the changes in species assemblages across the shelf perpendicular to the shore (depth assemblages). He firmly established that most species have discrete depth ranges and depths of maximum abundance. Major species assemblage boundaries occur on the northern shelf at 100 m, around 200 m, near 500 m, at 1000 m, and near 2000 m. Phleger also demonstrated that in marine marshes and in the upper reaches of the bays, lagoons, and estuary systems, agglutinant species are predominant. The low diversity and high standing populations of Mississippi Delta faunas were also carefully recorded.

Poag and Sweet (1972) summarized the foraminiferal studies of carbonate banks and reefs in the Gulf, showing that numerous submerged banks on the Texas-Louisiana and West Florida shelves are occupied by relict reefal foraminiferal assemblages of late Pleistocene age, whose composition varies significantly from living reef assemblages in the southern Gulf.

Since Phleger's (1960c) synthesis, six notable compendiums of benthic foraminiferal ecology and distribution have appeared. Walton (1964a) dealt extensively (and exclusively) with the northeastern Gulf of Mexico. Murray (1973) published a more generalized text that examines distributions on a global scale. Murray devoted only three and one-half pages to the Gulf, restricting his discussion to the northern and central Texas regions. Pflum and Frerichs (1976) provided additional collections of deep-water foraminifera from the northern Gulf. Their interpretations were attuned to depth distributions, and they suggested several additional species assemblages whose boundaries they associated with hydrostatic pressure changes.

Dignes (1978) used Q-mode factor analysis to recognize benthic foraminiferal species assemblages in the Gulf of Mexico, especially concentrating on the abyssal community. He related assemblage distributions to dissolved oxygen content, sediment type and composition, calcium carbonate and terrestrial organic carbon content, temperature, salinity, and water depth.

Douglas (1979) thoroughly reviewed the concepts and analytical methods used by benthic foraminiferal ecologists and paleoecologists, criticized earlier methods, and offered new data and interpretations. He dealt with Gulf of Mexico foraminiferal communities only in brief, general terms. Douglas and Woodruff (1981) have meticulously synthesized, on a global scale, knowledge of the abyssal benthic foraminifera and have provided a modern viewpoint of abyssal ecology and paleoecology.

Most of these authors have emphasized that distribution patterns of modern benthic foraminiferal species can be used to identify indirectly or monitor environmental properties and thereby can provide models for reconstructing ancient sedimentary environments. Other investigators have offered bountiful evidence that paleoenvironmental analyses based on foraminiferal species are eminently applicable to solving more complex geologic problems (for example, Berggren and Haq, 1976; Berggren et al., 1976). Walton (1964a), on the other hand, approached foraminiferal ecology and paleoecology on a higher taxonomic level, that of the genus.

I have adopted Walton's (1964a) analytical approach for this synthesis, for many of the same reasons that Walton expressed. He vividly demonstrated that the distributions of predominant genera in modern benthic foraminiferal assemblages delineate facies whose equivalents may be easily recognized in subsurface beds of Cenozoic age in the Gulf of Mexico region or elsewhere. Later, Sliter and Baker (1972) demonstrated close resemblances between the distribution of modern genera and that of Cretaceous homeomorphs. An analysis of the distribution of predominant genera is simpler and more easily achieved than an analysis of species distribution, and is especially useful in increasingly older rocks, as species assemblages differ increasingly more from modern analogs than do generic assemblages. I have modified Walton's techniques somewhat and applied them to the entire Gulf of Mexico benthic foraminiferal community. *Predominant genus* here means the genus in a given sample that constitutes the most abundant specimens in an aliquot of approximately three hundred specimens (or in the total assemblage if fewer than three hundred specimens constitute an entire sample) that have been retained on a 74 μm wet screen. (Some authors, including Walton, 1964a, would prefer the term *dominant* rather than *predominant*, but *dominance* implies more than merely greatest abundance; see Parker, 1975.) There is some variability in this counting method among published reports, but this variability does not significantly affect the generic distribution patterns. I refer to the benthic foraminiferal assemblages characterized by the predominance of a particular genus as generic predominance facies (for example, *Ammonia* predominance facies). In some cases, two or three genera are nearly equally abundant in a group of samples and the predominance facies delineated by those genera are labeled accordingly (for example, *Brizalina-Uvigerina* predominance facies). The generic predominance is calculated from total assemblages and would not necessarily be the same if only living specimens were used.

There are some exceptions to the use of predominant genera to define the distributional facies. The miliolids are lumped together as a group to identify a miliolid predominance facies because of the difficulty of separating miliolid genera consistently and because, as the distribution data show, the general environmental requirements for low latitude miliolids appear to apply rather uniformly to nearly the whole group (except for a few deep-water species). I have also grouped *Peneroplis* and *Archaias* together as soritids. Additionally, although the bathyal and abyssal agglutinant assemblages are quite distinct, no widespread pattern of a single predominant agglutinant genus has been detected among them. I have designated these agglutinant assemblages as the *Glomospira* predominance facies, although *Glomospira* is not necessarily predominant in every case. *Glomospira* is, however, one of the most consistently present forms, and it is often predominant. The distribution of the forty-two predominance facies is illustrated in Figures 8 to 19.

BAYS, LAGOONS, AND ESTUARIES

Charlotte Harbor

Bandy (1956) sampled Charlotte Harbor and adjacent waters. The *Ammonia* predominance facies occupies all harbor stations and extends into the lower reaches of the Peace and Caloosahatchee rivers, where it is

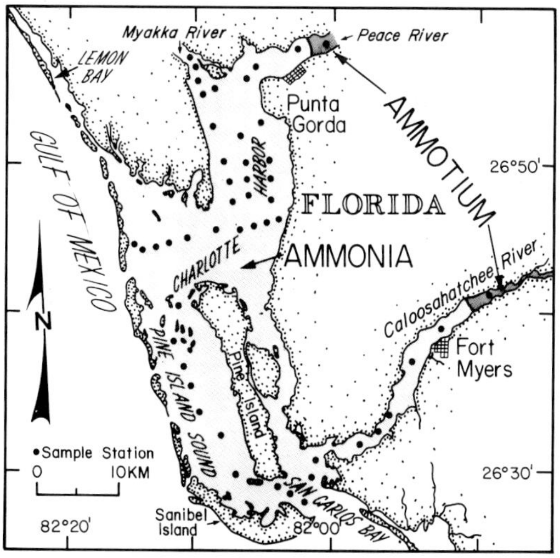

Figure 8
Generic predominance facies in Charlotte Harbor and environs, Florida (data from Bandy, 1956).

replaced upstream by the agglutinant *Ammotium* predominance facies. This facies pattern is repeated, with few modifications, throughout most of the other bays, lagoons, and estuaries around the Gulf (Figure 8).

Tampa, Hillsborough, and Sarasota Bays

Walton (1964b) investigated the foraminiferal ecology of these adjoining west Florida bays. The facies pattern is similar to that of Charlotte Harbor, except that the *Ammonia* predominance facies does not extend into the rivers, and small patches of the *Elphidium* and Miliolid predominance facies occur in the lower part of Tampa Bay (Figure 9).

Mobile Bay

In Mobile Bay, Alabama, the assemblages described by Walton (1964a) and Lamb (1972; Figure 10) are quite similar to those of the Tampa-Sarasota Bay complex. The difference is that the upper Mobile Bay region supports an agglutinant *Miliammina* predominance facies rather than that of *Ammotium*. Lamb (1972) reported that the middle and lower bay contains an *Ammonia* pre-

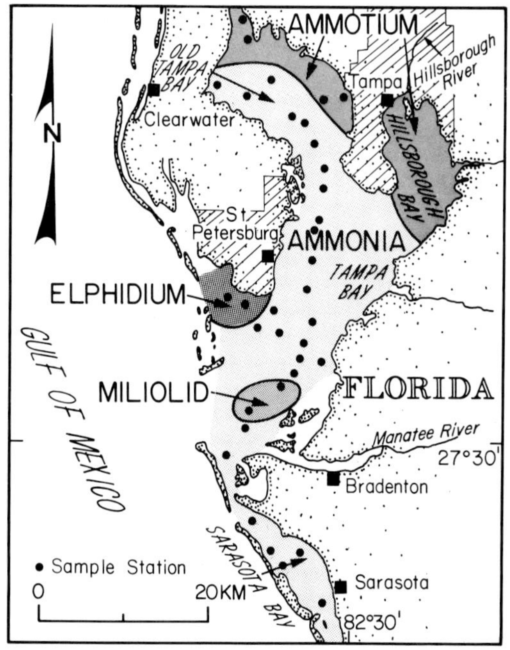

Figure 9
Generic predominance facies in Tampa, Hillsborough, and Sarasota bays, Florida (data from Walton, 1964b).

dominance facies, although Walton (1964a) recorded an *Ammotium* predominance facies in the middle of the bay. Walton called it *Ammobaculites*.

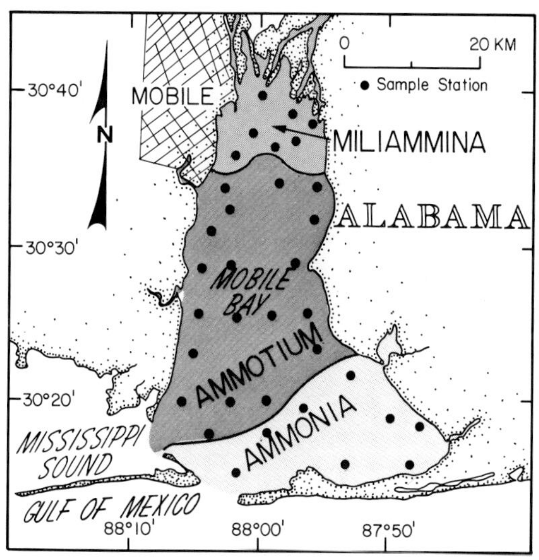

Figure 10
Generic predominance facies in Mobile Bay, Alabama (data from Lamb, 1972).

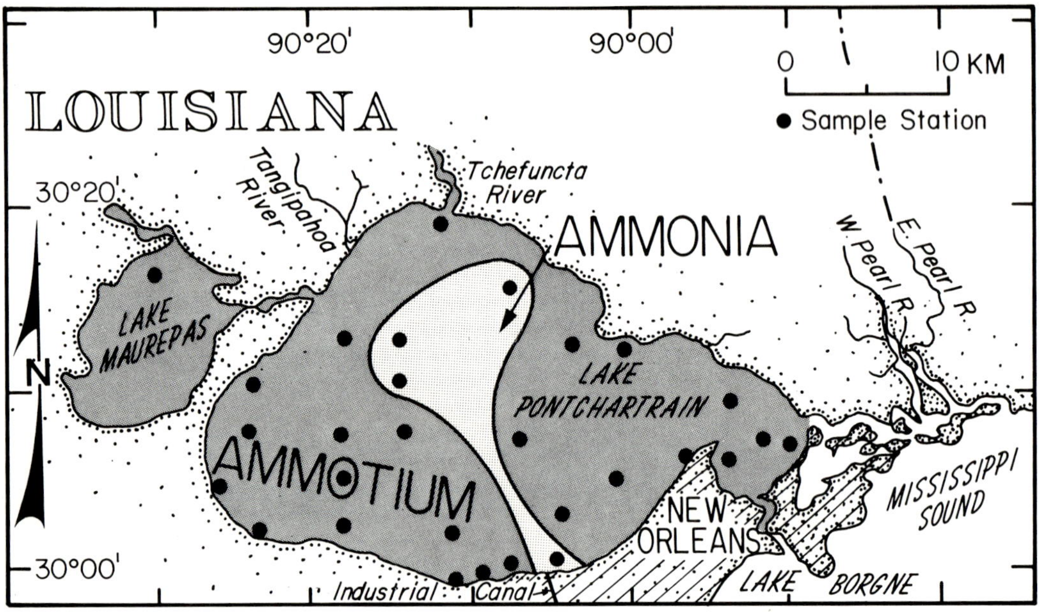

Figure 11
Generic predominance facies in Lakes Pontchartrain and Maurepas, Louisiana (data from Otvos, 1978).

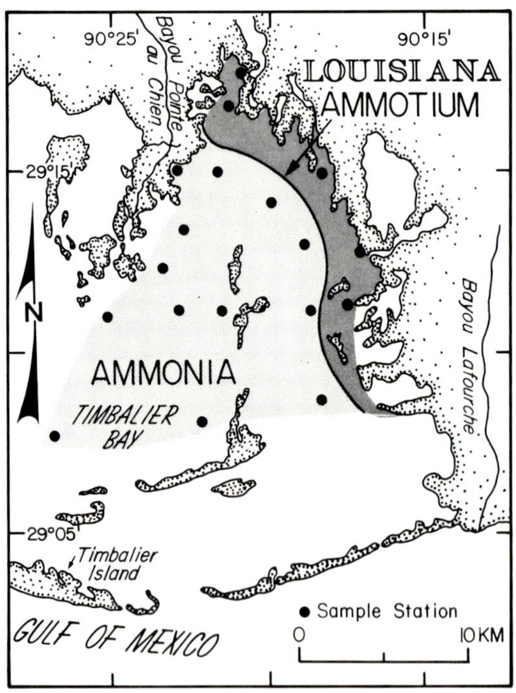

Figure 12
Generic predominance facies in Timbalier Bay, Louisiana (data from Waldron, 1963).

Lakes Pontchartrain and Maurepas

Otvos (1978) collected a series of samples from these interconnected brackish Louisiana lakes. The *Ammotium* predominance facies occupies both lakes except for a few stations in the center and southern shore of Lake Pontchartrain (Figure 11). The latter stations are influenced by saline Gulf water, which enters the lake by a dredged industrial canal, and the *Ammonia* predominance facies is developed there.

Timbalier Bay

In Timbalier Bay, Louisiana, Waldron's data (1963) reveal the *Ammotium* predominance facies adjacent to the shoreline marshes and the *Ammonia* predominance facies throughout the rest of the bay (Figure 12).

Sabine Lake

Sabine Lake on the Texas-Louisiana border is more isolated from the Gulf than the other estuaries discussed here, but the typical upper bay *Ammotium* predom-

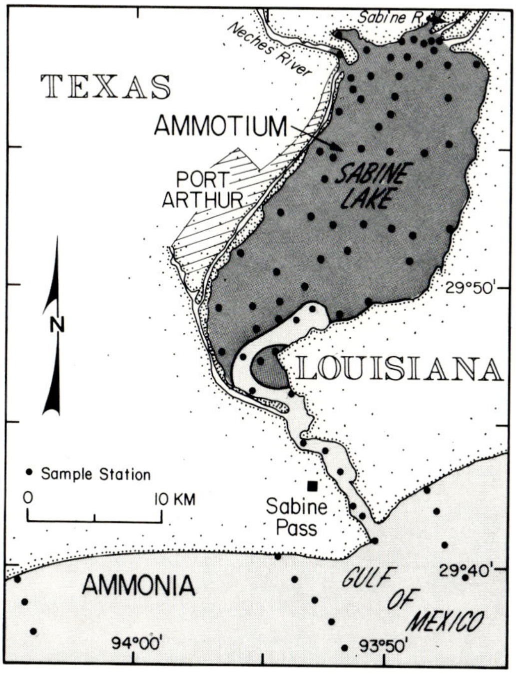

Figure 13
Generic predominance facies in Sabine Lake and environs, Texas-Louisiana (data from Kane, 1962, 1967).

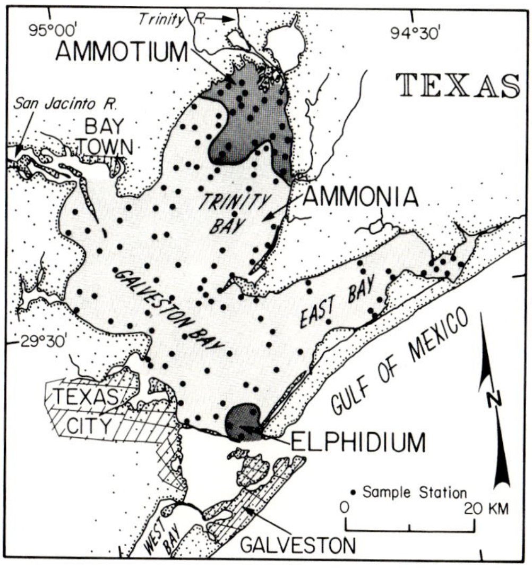

Figure 14
Generic predominance facies in Trinity and Galveston bays, Texas (data from Wantland, 1969; W. V. Sliter, written comm., 1978).

inance facies is clearly defined (Figure 13; Kane [1967] called these forms *Ammobaculites*). The lower southern part of Sabine Lake and its connecting channel contain the *Ammonia* predominance facies.

Trinity and Galveston Bays

In Trinity Bay, Texas, Wantland (1969) recorded the *Ammotium* predominance facies adjacent to the Trinity Delta (Figure 14). The *Ammonia* predominance facies occupies the lower part of Trinity Bay and most of Galveston Bay. The *Elphidium* predominance facies is present at the mouth of Galveston Bay (W. V. Sliter, U.S. Geological Survey, written communication).

San Antonio Bay

Data from Parker, Phleger, and Peirson (1953) and Poag (1976, 1978) show that the *Ammotium* predominance facies is present around the Guadalupe River delta in the upper part of San Antonio Bay, Texas (Figure 15). A narrow finger of this facies extends about a third of the distance down the center of the bay following the outflow of river water toward the Gulf. The rest of the bay contains the *Ammonia* predominance facies.

Baffin Bay and Laguna Madre

This south Texas area contains the major exception to the patterns described above (Figure 16). Baffin Bay contains the *Ammonia* predominance facies as expected, but apparently it has not been sampled far enough upstream to reach the *Ammotium* predominance facies. In most of the Texas part of Laguna Madre, however, the *Ammonia* predominance facies is replaced by the Miliolid predominance facies. A narrow strip of the *Elphidium* predominance facies is also present. The Tamaulipas part of the laguna, on the other hand, does contain chiefly the expected *Ammonia* predominance facies (Phleger, 1960a, b).

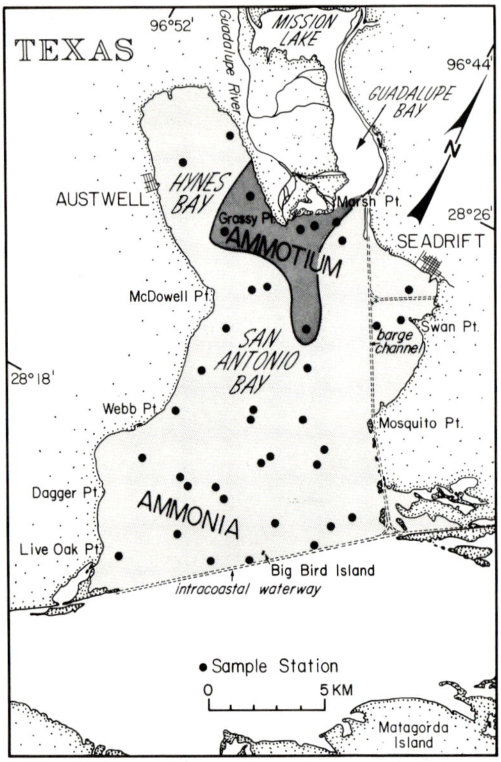

Figure 15
Generic predominance facies in San Antonio Bay (data from Poag, 1976).

Lagunas Camaronera and Alvarado

Phleger and Lankford (1978) recorded the *Ammotium* predominance facies in all samples taken from Laguna Camaronera, in the upper reaches of Laguna Alvarado, and along the southwestern shore of Laguna Alvarado along the Veracruz coast adjacent to the mouths of the Rio Blanco and Rio Papaloapan (Figure 17). The *Ammonia* predominance facies occupies the center and northeastern margin of Laguna Alvarado.

Laguna de Terminos

In this southernmost major estuary of the Gulf, the *Ammonia* predominance facies is still widespread (Figure 18). Near the river mouths, however, Ayala-Castañares (1963) recorded an *Elphidium* predominance facies rather than an *Ammotium* predominance facies. Presumably samples were not collected far enough upstream to encounter the *Ammotium* predominance facies (in the Charlotte Harbor area, the *Ammotium* predominance facies occurs 30 km up the Caloosahatchee River).

To summarize, the *Ammonia* predominance facies is the most consistent and widespread facies in the main bodies of coastal bays, lagoons, and estuaries, but *Ammonia* is generally supplanted by the *Ammotium* predominance facies near river mouths. The Miliolid and *Elphidium* predominance facies also are found scattered in these coastal waters.

Figure 16
Generic predominance facies in Laguna Madre and lower Baffin Bay, Texas (data from Phleger, 1960b).

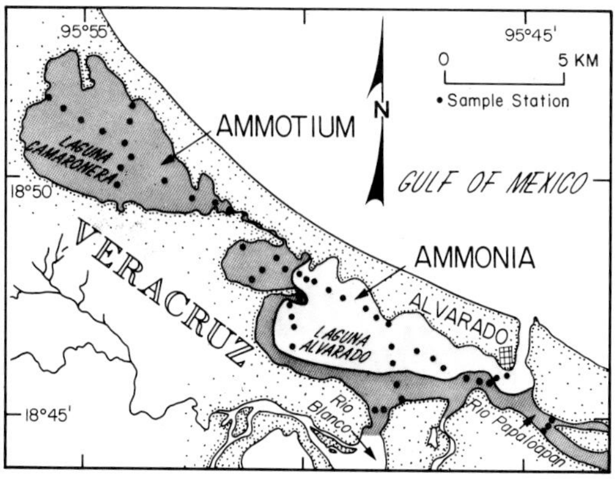

Figure 17
Generic predominance facies in Lagunas Camaronera and Alvarado, Veracruz (data from Phleger and Lankford, 1978).

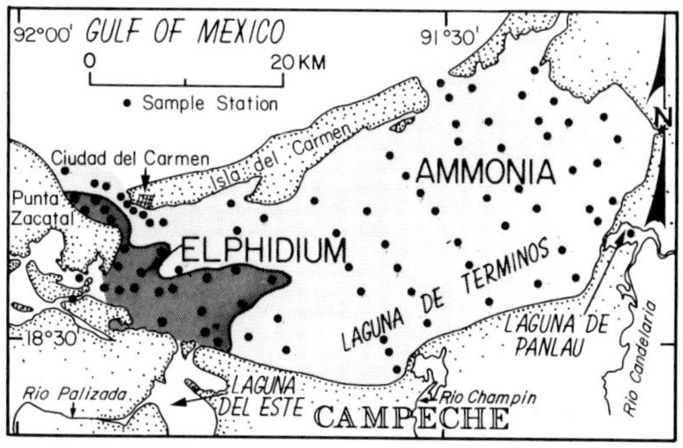

Figure 18
Generic predominance facies in Laguna de Terminos, Campeche (data from Ayala-Castañares, 1963).

CONTINENTAL SHELF

The distribution patterns of generic predominance facies on the continental shelf are much more complex (Figure 19, in pocket) than the simple ones in the estuaries. The general pattern is a mosaic of numerous elongate coastwise predominance facies. Two predominance facies nearly encircle the Gulf, but the others are more provincial.

Next to the shore, the *Ammonia* predominance facies extends from the lower parts of the estuaries into the inner sublittoral realm of the open Gulf. This is the most widespread of all predominance facies on the shelf. The *Ammonia* predominance facies is continuous from the tip of Florida to Laguna de Terminos, except for a brief interruption at the Mississippi Delta. It is somewhat modified, however, along the Texas and Mexican shelves south of Galveston Bay, where *Ammonia* shares its predominance nearly equally with *Elphidium* and forms the *Ammonia-Elphidium* predominance facies.

Farther out on the northern shelves is a diverse patchwork of locally restricted predominance facies. The predominance facies surrounding the Mississippi Delta, for example, have no known counterparts elsewhere in the Gulf. Adjacent to the delta, assemblages immensely rich in specimens are characterized by the small, fragile-walled genera *Epistominella* and *Nonionella*. In fact, each of these generic groups is nearly monospecific here. The most basinward predominance facies of the Mississippi Delta is even more unusual. The agglutinant genus *"Goesella"* predominates in the form of a single species, *"Goesella" mississippiensis* Parker. This is one of the few predominance facies whose nominate taxon is indigenous to the Gulf. Furthermore, it is restricted to the Mississippi Delta area.

Another restricted agglutinated predominance facies is found south of the Mississippi Delta between the *Nonionella* and *"Goesella"* predominance facies; this is characterized by the genus *Nouria*. As in the other predominance facies around the delta, the predominant genus is monospecific, composed solely of *Nouria polymorphinoides* Heron-Allen and Earland.

Separated by the Mississippi Delta, the groups of predominance facies of the West Florida Shelf are quite different from those of the Texas-Louisiana Shelf. In fact, only four predominance facies (*Ammonia*, *Cibicidoides*, *Amphistegina*, and *Brizalina*) are common to these two shelf regions. Aside from *Ammonia*, the most widespread of the seventeen West Florida Shelf predominance facies is the Miliolid predominance facies. This is most strongly developed off southern Florida and extends parallel to and just basinward of the *Ammonia* predominance facies as far north as Cape San Blas. An isolated strip of Miliolid predominance facies occurs near the shelf edge off Cape San Blas; this may be a relict assemblage. Ten other predominance facies fill the rest of the shelf off peninsular Florida (Figure 19, in pocket). The mid-shelf predominance facies are characterized by *Asterigerina, Rosalina,* and several soritids. *Hanzawaia* and *Planulina* predominance facies lie between these and the shelf-edge *Cibicidoides* predominance facies, and the *Amphistegina* predominance facies occupies scattered submerged banks.

On the shelf off panhandle Florida, Mississippi, and Alabama, *Amphistegina* and *Planulina* predominance facies occur along with four totally different predominance facies. *Rosalina, Rosalina-Hanzawaia,* and *Islandiella-Rosalina* constitute the predominance facies of the middle and outer shelf, and *Brizalina* replaces *Cibicidoides* as the shelf-edge predominance facies.

Seaward of the *Ammonia* and *Ammonia-Elphidium* predominance facies, the Texas-Louisiana Shelf contains ten predominance facies. *Fursenkoina, Saccammina-Ammobaculites, Elphidium-Hanzawaia, Buliminella, Bigenerina,* and *Elphidium* characterize the middle shelf, whereas *Brizalina, Cibicidoides,* and, locally, *Uvigerina* occupy the outer shelf. *Amphistegina* is concentrated on submarine banks.

The narrow shelf off eastern Mexico is poorly sampled, but the *Elphidium* predominance facies can be recognized on the middle shelf, and *Brizalina* and *Brizalina-Uvigerina* predominance facies occupy the shelf edge. The Miliolid-*Archaias-Homotrema* predominance facies occupies scattered reefs.

The Campeche Shelf is an enormous expanse of the Miliolid predominance facies. This predominance facies is bordered at its outer edge by the *Cibicidoides* predominance facies as it is off Florida. Emergent reefs are characterized by the Miliolid-*Archaias-Homotrema* predominance facies.

CONTINENTAL SLOPE

On the continental slope the predominance facies resume their circum-Gulf patterns (Figure 19, in pocket). The upper slope is clearly recognized by the preponderant *Brizalina* predominance facies. This facies is interrupted only off the Mississippi Delta, where *Bulimina* is predominant, and at the Yucatán Channel and the Straits of Florida, where data are unavailable. The *Brizalina* predominance facies is modified off Mexico, however, where *Brizalina* and *Uvigerina* are nearly equally abundant on the upper slope and form the *Brizalina-Uvigerina* predominance facies. Also, the *Uvigerina* predominance facies occupies a narrow strip within the *Brizalina* predominance facies on the Texas-Louisiana Slope.

The lower continental slope predominance facies is that of *Bulimina*. It, too, is nearly continuous around the Gulf, but four interruptions occur. *Islandiella* replaces it off Cape San Blas and off southern Florida, *Brizalina* dilutes it off central Texas, a patch of *Cibicidoides* occupies the middle of the wide Campeche Slope, and the steep walls of the Campeche Canyon mask the *Bulimina* predominance facies off northwestern Yucatán.

CONTINENTAL RISE AND FLORIDA PLAIN

The upper part of the continental rise is typified by an assemblage of minute calcareous species, predominated by *Nuttallides decorata* (Phleger and Parker)(Figure 19, in pocket). This *Nuttallides* predominance facies is continuous around the Gulf, crosses the upper part of the Mississippi Fan, and even borders the southeastern corner between Yucatán and south Florida where it covers the Florida Plain. A small patch of the *Eponides* predominance facies nearly replaces the *Nuttallides* predominance facies in the DeSoto Canyon area, and local patches of the *Hoeglundina* and *Glomospira* predominance facies occur within the *Nuttallides* predominance facies in the eastern and western Gulf, respectively.

SIGSBEE PLAIN AND MISSISSIPPI FAN

The deepest part of the Gulf of Mexico contains two easily recognized predominance facies (Figure 19, in pocket). The *Cibicides* predominance facies, characterized nearly exclusively by the large species *Cibicides wuellerstorfi* (Schwager), occupies the bulk of the Mississippi Fan, whereas the Sigsbee Plain and the adjacent lower part of the Western Gulf Rise contain mainly minute forms of *Eponides* and *Nuttallides*. However, *Eponides* tends to be slightly more abundant than *Nuttallides* at some stations on the Western Gulf Rise. Closer spaced samples might allow delineation of a separate *Eponides* predominance facies on this segment of the rise.

The remaining deep-water predominance facies is the widely scattered *Glomospira* predominance facies. In it, agglutinants such as *Saccorhiza, Adercotryma,* and *Glomospira* are predominant, and calcareous forms are scarce. The *Glomospira* predominance facies usually occurs in isolated samples; therefore the size of the patches representing this facies in Figure 19 is exaggerated.

RELATIONSHIPS BETWEEN GENERIC PREDOMINANCE FACIES AND ENVIRONMENTAL PROPERTIES

A host of environmental factors are known to influence the distribution of living benthic foraminifera. Murray's (1973) comprehensive review, for example, dealt with temperature, salinity, availability of calcium carbonate, water depth, pH, oxygen, illumination, and food. Phleger (1960c) included turbidity, type of substrate, biologic competition, the effects of water masses, and hydrostatic pressure. Greiner (1974) emphasized the interrelationships of coastal physiography and climate and the availability of calcium carbonate. Field and culture studies have demonstrated a clear global correlation between foraminiferal distribution and salinity, but direct evidence for the importance of the other factors is more limited.

We, of course, are not dealing here with living populations. The distributions I (and most other authors) have described for the Gulf of Mexico are based on total benthic foraminiferal faunas, which include chiefly tests of dead specimens. This is an appropriate approach for paleoecologists, for this is what they see in fossil assemblages. As a result of including dead assemblages, however, some additional environmental factors must be accounted for. A major geologic influence on the patterns exhibited is the effect of lowered sea level during the last advance of Pleistocene continental ice sheets. This resulted in migration of foraminiferal predominance facies across the continental shelf and slope as the shoreline advanced basinward (Poag and Valentine, 1976). Lowered sea level also caused frequent mechanical displacement of shallow-water assemblages into deeper water as turbidity flows and shelf-edge deltas spread sublittoral and fluvial sediments across the outer shelf and down the slope (Sidner, Gartner, and Bryant, 1979). The chemistry of water masses (Kennett and Schackleton, 1975) and sedimentary deposition (Thunnell, 1976; Hedges and Parker, 1976; Dignes, 1978) were also considerably altered, and the changes may be still influential in some regions.

The scope of this investigation precludes consideration of many of the environmental factors enumerated above because the pertinent data either are not available or are known chiefly from local estuarine systems. However, there are some obvious correlations between the distribution of generic predominance facies and several specific environmental properties. Readers should keep in mind that correlation with a certain environmental property does not necessarily imply control by that property. Unknown (or unmeasured) covariant properties (or groups of properties) may exert equal or dominant influence that ultimately controls the facies distributions.

SALINITY

Most investigators agree that salinity exerts the most evident and perhaps the dominant influence on the estuarine and near-shore foraminifera. This is because the highly variable diurnal, seasonal, and annual salinity fluctuations demand strongly euryhaline species and genera, and there are not many among the foraminifera. The present synthesis emphasizes the substantial influence of salinity in estuarine areas. Nearly all of the estuaries contain the agglutinated *Ammotium* predominance facies in the river-freshened upper reaches, and this facies is typically replaced in the more saline lower parts of estuaries by the *Ammonia* predominance facies (Figures 8 to 18). Among those studied, the only major coastal lagoon whose lower part does not contain chiefly the *Ammonia* predominance facies is the Texas portion of Laguna Madre. Rusnak (1960) showed that the Texas part of Laguna Madre differs from other Gulf-margin estuaries and lagoons in having a turbid, hypersaline water mass caused by limited freshwater inflow and a semiarid climate. For geological purposes, Rusnak considered this part of Laguna Madre as an enclosed lagoon. Phleger (1960a,b) attributed the presence of the Miliolid predominance facies here to the hypersalinity of the water and perhaps to the sandier substrate. Murray's (1973) work has shown that some miliolids cannot maintain their pseudopodial feeding reticulum in low salinity water, so the high salinity may contribute to their abundance. *Ammonia* can tolerate high salinities also, however, as well as sandy substrates, so the cause of this Miliolid anomaly appears to be more complex.

The effect of salinity may be further demonstrated by the interruption of the *Ammonia* predominance facies off the Mississippi Delta and on the Campeche Shelf (Figure 19, in pocket). In the Mississippi Delta, the freshwater discharge is enormous, whereas off Yucatán, there is (as in the Texas part of Laguna Madre) virtually no freshwater runoff (Logan, 1969). Although *Ammonia* is demonstrably euryhaline, apparently it does not compete successfully with other more specialized genera in these two ecologically extreme areas.

High salinity also contributes significantly to the broad distribution of the Miliolid predominance facies on the West Florida and Campeche shelves (Figure 19, in pocket), but other factors may be just as important. Both of these areas receive very little terrigenous detritus from rivers. These shelves are overlaid by shallow, warm, clear waters that are penetrated deeply by sunlight and that provide ideal biotopes for lush growth of sea plants. Murray (1963), Brasier (1975a), and others have demonstrated the close relationship between the epiphytic miliolids and marine vegetation. Also, these two major expanses of the Miliolid predominance facies occur where calcium carbonate, apparently produced mainly by calcareous algae, is abundant on the sea floor.

TEMPERATURE

The effect of temperature is most readily shown by the close correlation of the Miliolid predominance facies with the regions of negligible annual temperature variation. A line across the Gulf that separates the Miliolid predominance facies from the mosaic patterns of the northern shelves is nearly coincident with the boundary between the Carolinian and Caribbean biogeographic provinces (Figure 2), which is also largely determined by latitudinal temperature patterns.

Theoretically one would expect a correlation between bathymetric temperature changes and the change in predominance facies across the continental slope, rise, and abyssal plains. Lohman (1978) has shown close correlation between bottom-water temperature and benthic assemblage changes in the South Atlantic, for example. The deep-water temperature data are too generalized at present, however, to delineate any definite correlations with predominance facies trends in the Gulf of Mexico.

SUBSTRATE

The distribution of sediments in the Gulf of Mexico is intimately related to the physiographic features of the Gulf and surrounding coastal plain. In turn, we can see a close correspondence between the generic predominance facies patterns and the types of substrates with which they are associated.

On the shelves covered by siliciclastic sediments are found numerous coastwise bands of predominance facies (Figure 19, in pocket). On the other hand, in the two large tracts of carbonate sediments, a single assemblage, the Miliolid predominance facies, is present. Furthermore, the West Florida Shelf, whose sediments are often quartz sand mixed with biogenic carbonate, contains predominance facies that are different from those of the Texas-Louisiana shelf, which contains more terrigenous silt, clay, and organic debris and has very little carbonate.

The sediments of the upper slope, which contain maximum concentrations of organic carbon (Trask, 1953), are occupied by predominance facies whose nominate genera, *Brizalina* and *Uvigerina*, are known to abound in low-oxygen environments (Douglas, 1979). Oxidation of the organic debris may produce an oxygen deficiency at the sediment-water interface, which could account, in part, for this distribution pattern.

Two broadly spread predominance facies are present in the deep Gulf basin. They apparently have no relationship to water mass, as a single water mass is reported to cover both of them. (Perhaps subtle differences are as yet undetected in the water masses.) There is a clear correlation, however, between the distribution of the two predominance facies and the sediment type (Figure 19, in pocket). The *Cibicides* predominance facies is restricted to the area covered by the bulk of the Mississippi Fan, an area where sediments are rapidly deposited (hence the fan shape) and are especially rich in terrestrial organic detritus (Huang and Goodell, 1970; Koons and Perry, 1976; Hedges and Parker, 1976; Dignes, 1978). The *Eponides-Nuttallides* predominance facies is restricted to the Western Gulf Rise and Sigsbee Plain, where the sedimentation rate is much lower, and the organic component of the sediments is diminished and of generally more marine origin. Any or all of these factors might influence the benthic foraminiferal populations (see Dignes, 1978).

WATER MASS COMPOSITION

As Phleger and Parker (1951) have pointed out, the northern shelves down to about 100 m depth are overlaid by the well-mixed seasonal water mass (Figure 5). This seasonal mass also encompasses most of the shelf photic zone and the area most affected by the turbulent forces of longshore currents, tides, and waves. Precisely beneath this seasonally changing water mass, the most complex patterns of coastwise predominance facies occur (out to the inner edge of the *Brizalina* predominance facies Figure 19, in pocket). In contrast, on the southern shelves (Campeche and southern Florida), where seasonal changes in the shelf water mass are small, the facies pattern is broad and simple (in fact, a single predominance facies).

Between the seasonally mixed surface layer (∿100 m) and the environmentally most stable waters beneath the permanent thermocline (>1500 m), salinity, temperature, and dissolved oxygen are constant on a yearly basis at any given depth, but they vary considerably at different depths. Where these four distinct water masses impinge upon the floor of the Gulf, the nearly circum-Gulf predominance facies of the outermost shelf and the slope are developed: those of *Brizalina, Brizalina-Uvigerina, Bulimina,* and *Cibicidoides*. At present, no direct correlation is obvious between any one of these predominance facies and a particular water mass, but there is indirect evidence that the *Brizalina-Uvigerina* and the *Brizalina* predominance facies may be associated, in part, with the oxygen minimum layer, which occupies the depths between 300 m and 900 m (Dignes, 1978; Douglas, 1979).

The deepest water mass, the Gulf Basin Water, covers the entire remaining basin floor below 1500 m and maintains presumed constant salinity, temperature, and dissolved oxygen. Around the upper rim of this water mass, the *Nuttallides* predominance facies is developed around the entire Gulf, and below that lie the *Eponides-Nuttallides* and *Cibicides* predominance facies.

The upper boundary of the *Nuttallides* predominance facies coincides approximately with the sill depth across the Yucatán Channel, suggesting a correlation with possible subtle changes in the Gulf Basin Water associated with the sill depth. The distinct east-west division of the deepest predominance facies also suggests subtle compositional differences within the water mass between these two regions, but none has been yet noted.

BOTTOM TOPOGRAPHY

The deeper parts of the Gulf slopes and rises are notably hummocky because of the movement of underlying ridges and diapirs of salt and shale. This is especially true of the slope off Louisiana, Texas, Tamaulipas, and Campeche (Lehner, 1969; Bryant et al., 1968; Martin and Bouma, 1979). Plotting the patches of agglutinant *Glomospira* predominance facies on a more detailed bathymetric map (see Sorenson et al., 1975) shows clearly that they are usually associated with bottom prominences (Figure 19, in pocket). A detailed study around one such prominence has revealed that the Holocene sediments are much thinner here than in the adjacent basin (Bouma et al., 1975; Sidner and McKee, 1976; Sidner, Gartner and Bryant, 1979). The sediments are also highly oxidized, containing large quantities of iron and manganese both as free constituents in the sediments and as replacement and fillings in the planktic foraminiferal tests. Some of the agglutinant forms that use collages of calcareous planktic tests to form their chambers show significant dissolution of the calcite, which leaves large cavities in the surrounding siliceous/organic matrix of the test wall (C. W. Poag, unpub. data). Even the most dissolution-resistant planktic tests are eroded and fragmented. Apparently the long exposure to the bottom water due to low sedimentation rate on the prominences has altered the original faunal composition by depleting the carbonate tests and thereby enriching the samples in agglutinant species. Thus the *Glomospira* predominance facies is primarily a product of the unusual geochemical environment on the deeply submerged ridges and knolls.

The relative abundance of *Nuttallides* and *Eponides* on the Western Gulf Rise may also be controlled, in part, by calcite dissolution. Dignes (1978) found a strong correlation between the distribution of abundant *N. decorata* and low carbonate dissolution; *Eponides turgidus* Phleger and Parker showed the reverse relationship.

HOLOCENE SEA LEVEL RISE

During the last Pleistocene glacial interval, sea level in the Gulf of Mexico was considerably lower than at present, perhaps as much as 150 m (Poag, 1973). At this time, estuarine, deltaic, and inner shelf biotopes were present at or near what is now the shelf edge. In some areas, such as off Texas, slow or negligible modern sedimentation has left parts of these earlier deposits, including relict foraminifera, still exposed at the sea floor (Shepard et al., 1960). The most obvious occurrence of these relict Pleistocene foraminifera is on the submerged banks and ridges along the Texas-Louisiana and West Florida Shelf edge (Figure 20). In addition, the distribution of *Elphidium, Elphidium-Hanzawaia*, and *Bigenerina* predominance facies (Figure 19, in pocket) approximately coincide with a large patch of relict sediment noted by Shepard et al. (1960) on the Texas-Louisiana Shelf. Also, the *Planulina* and *Cibicidoides* predominance facies of the outer West Florida and Campeche shelves

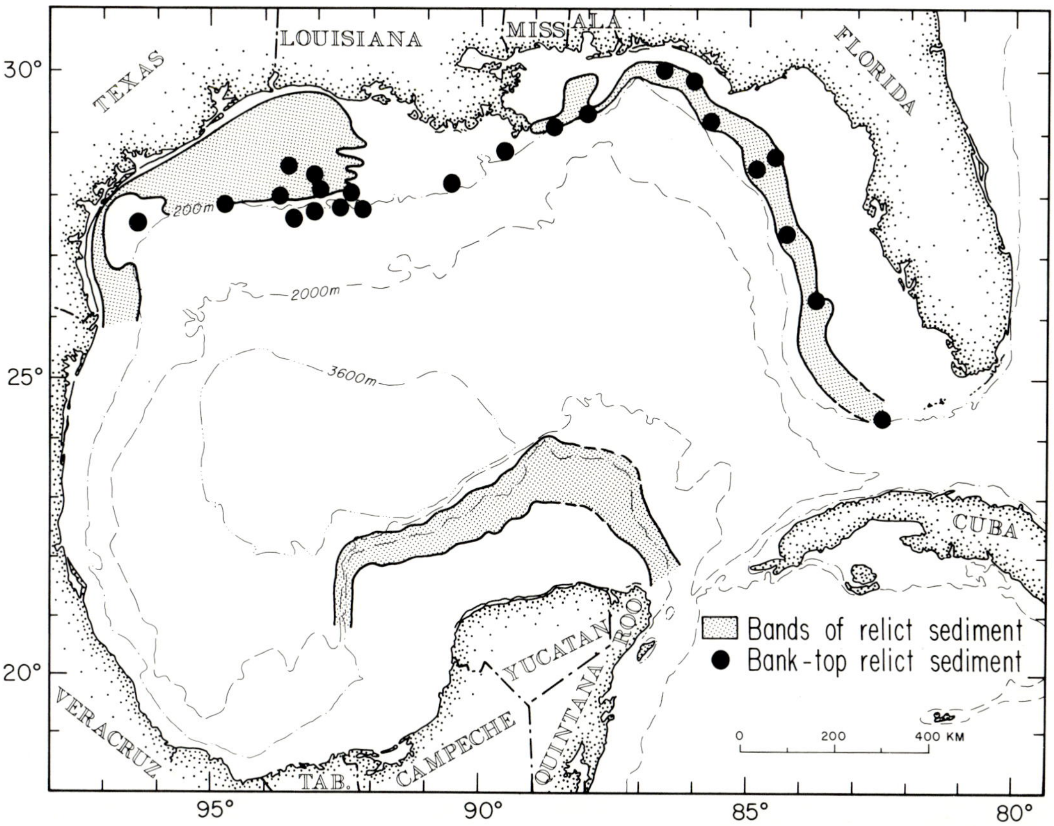

Figure 20
Distribution of known relict sediments and foraminiferal assemblages in Gulf of Mexico (data from Gould and Stewart, 1955; Bandy, 1956; Shepard et al., 1960; Logan et al., 1969).

(Figure 19, in pocket) closely coincide with bands of relict algal and oolitic sands (Gould and Stewart, 1955); Bandy, 1956; Logan et al., 1959). These relationships suggest that these five generic predominance facies may be the result of premodern accumulations.

There are, in addition, some submerged banks whose flanks were eroded during the low Pleistocene sea level stands, thus exposing Tertiary strata. The encompassed Tertiary microfossils are still being incorporated into the modern assemblages (Poag and Sweet, 1972). The Pleistocene strata that currently blanket most of the coastal margin of the Gulf and that covered the exposed shelf during the last Pleistocene glacial stage have prevented wholesale redeposition of Tertiary microfossils into the modern coastal assemblages. But locally, around river mouths, fossil foraminifera as old as late Cretaceous age have been transported from inland outcrops and have been mixed into the modern nearshore sediments (Poag, 1972).

GENERIC PREDOMINANCE DIVERSITY

In the last few years, a number of deductive models have been introduced that assist in comprehending the relationships between specific aspects of community organization and structure (such as species diversity, species equitability, spatial dispersion, morphologic variability, and rates of evolution) and properties of the environment (such as physicochemical stability, physicochemical rigor, trophic structure, nutrient availability, spatial heterogeneity, and habitat disturbance (Sanders, 1968; Levins, 1968; Bretsky and Lorenz, 1970; Levinton, 1970, 1973, 1974; Grassle and Sanders, 1973; Somero and Soulé, 1974; and Bernstein et al., 1978). Some of these models have been tested among various groups of living and fossil invertebrates and meiofauna (for example, Jackson, 1972, 1973; Ashton and Rowell, 1975; Kauffman, 1977; Bernstein et al., 1978; Pachut and Antsey, 1979).

Evidence is mounting that taxonomic diversity is positively correlated with the physicochemical stability and rigor and the biotic disturbance of the environment, although not all tests have yielded positive results (Ashton and Rowell, 1975; Osman and Whitlatch, 1978). Physicochemical properties appear to be especially effective regulators in coastal and continental shelf biotopes, but biological interactions may be more important in bathyal and abyssal biotopes (Sanders, 1968; Dayton and Hessler, 1972; Grassle and Sanders, 1973).

Three important papers have dealt with species diversity patterns among benthic foraminifera (Buzas and Gibson, 1969; Gibson and Buzas, 1973; Bernstein et al., 1978). Gibson and Buzas (1973) showed that foraminiferal species diversity consistently increases from the inner to outer shelf in twenty traverses in the western North Atlantic Ocean and the Gulf of Mexico. This shelf trend has since been substantiated in additional studies from other ocean basins (Sen Gupta, 1977) and is predicted by the popular stability-time hypothesis of Sanders (1968).

In bathyal and abyssal biotopes, however, the patterns of foraminiferal species diversity are not globally consistent. This is clearly evident when comparing species diversity trends in the Gulf of Mexico and the western North Atlantic (Gibson and Buzas, 1973). Species diversity is maximum in abyssal depths of the North Atlantic, whereas diversity decreases below sill depth ($\sim$2000 m) in the Gulf of Mexico. The reasons for these differences currently are not understood, but we can make some reasonable speculations. On the theoretical side, the stability-time hypothesis would predict highest diversities in abyssal depths as a result of long-term physicochemical stability of that environment. On the other hand, Kauffman (1977) has pointed out that abyssal communities should be highly adapted to the rigorous abyssal environment (such as extreme temperatures and pressures), and therefore should have undergone slow evolution, which, in turn, would result in low diversity. This concept of slow evolution is supported by the long fossil record of many deep-sea foraminifera (Douglas and Woodruff, 1980).

Better-known physicochemical mechanisms, such as preferential dissolution of calcium carbonate tests (variable carbonate saturation of deep water masses), may play a role in controlling some bathyal and abyssal diversity patterns (for example, Bremer et al., 1979). Dignes (1978) has suggested this possibility for Gulf of Mexico abyssal assemblages. Furthermore, small sample size and incomplete recovery of core tops, which characterize nearly all piston and gravity cores, can seriously bias species diversity toward low values (Douglas and Woodruff, 1980).

Perhaps of greatest significance is the isolation of deep-water masses in silled basins such as the Gulf of Mexico. For example, in the silled basins of the Arctic Ocean (Lagoe, 1978), the Norwegian-Greenland Sea (Belanger and Streeter, 1980), and the Mediterranean Sea (Parker, 1958), foraminiferal species diversity is low below sill depth, as in the Gulf of Mexico. Douglas and Heitman (1979) have documented similar low diversities in the smaller silled basins of the California Borderlands.

Surprisingly there have been no specific studies of modern foraminiferal diversity at the generic level. But the foraminiferal data from the Arctic Ocean (Lagoe, 1978), the Norwegian-Greenland Sea (Belanger and Streeter, 1980), and the Mediterranean Sea (Parker, 1958) clearly show that generic diversity trends parallel the species diversity trends in those silled basins. Figure 19 (in pocket) shows that trends of generic predominance diversity in the Gulf, defined as the number of different generic predominance facies in a given physiographic realm, mimic the trends in species diversity (Gibson and Buzas, 1973). Generic predominance diversity is low (six facies) in the estuaries and inner shelf biotopes, increases dramatically to twenty-nine facies on the middle and outer shelf, declines sharply to eight facies on the slope, and reaches a minimum below sill depth (five facies on the Mississippi Fan; three on the Sigsbee Plain).

FOSSIL FORAMINIFERA AND PALEOECOLOGICAL INTERPRETATION

The analytical technique of delineating the distribution of generic predominance facies lends itself readily to the paleoecological interpretation of fossil associations of benthic foraminifera. Such interpretations are especially applicable to the Cenozoic rocks of the Gulf of Mexico into which more than sixteen thousand wells have been drilled during the exploration for and production of petroleum and where assembly of paleoecological maps forms an important part of the exploration procedures (see Robinson and Kohl, 1978).

The thick Pleistocene deposits below the Texas-Louisiana Shelf are especially suited to this approach because most of the Pleistocene foraminiferal species and genera are still extant, and their environmental relationships are known better than those of older fossil taxa. Poag and Valentine (1976) demonstrated that the recognition of changing paleoenvironments through the last 2 million years provided a coherent history of cyclic deposition and subsidence within the Texas-Louisiana Pleistocene basin. By documenting the lateral migration of generic predominance facies through geologic time, they clearly showed that cycles of sea-level fluctuation were directly associated with paleotemperature changes (as inferred from other faunal evidence) during the advance and retreat of North American Pleistocene ice sheets (Figure 21).

The extension of these techniques back further through geologic time becomes increasingly less precise because the number of extinct fossil genera (and species) increases. However, the problem of extinct fossil genera is not really severe for Tertiary faunas (those developed during the last 65 million years) and, in fact, is negligible from the

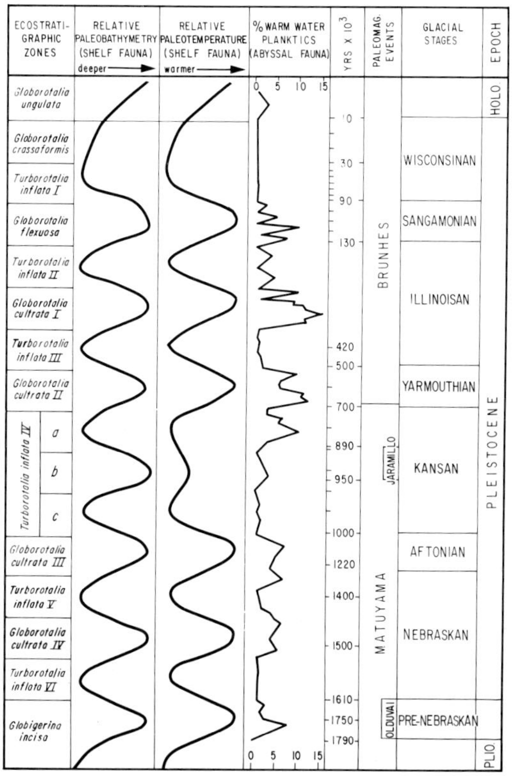

Figure 21
Example of how generic predominance facies can be used to reconstruct a curve of relative paleobathymetry (column 2), which in turn can be correlated with ecostratigraphic zones (column 1), relative paleotemperature (columns 3 and 4), and a radiometric and paleomagnetic time scale (columns 5 and 6) to reconstruct the geologic history of a sedimentary basin where subsidence and eustatic sea level fluctuations dominate tectonic movements (data from Poag and Valentine, 1976).

late Oligocene (30 million years ago) to present, because most of the later Tertiary genera are extant.

Several Gulf Coast investigators have demonstrated the applicability of the generic predominance method in Miocene and Oligocene sediments, but the best demonstration of its utility in middle Tertiary strata is the study by Walton (1964a). Walton used fossil generic predominance data from approximately 360 wells in the south Texas coastal plain to identify and correlate marine transgressions and regressions that took place during Oligocene time. He then was able to construct a map showing paleobathymetry of the Oligocene sea during deposition of the subsurface Anahuac Formation (Figure 22).

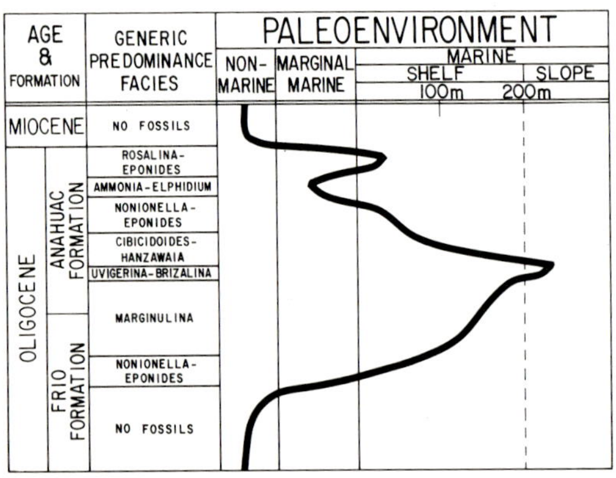

Figure 22
Paleoenvironments of subsurface Oligocene rocks in coastal Texas as determined from generic predominance facies (data from Walton, 1964a).

Poag (1980) used benthic generic predominance facies along with other faunal characteristics to interpret transgressive-regressive cycles of sediment deposition in rocks as old as Jurassic in the Baltimore Canyon Trough (offshore New Jersey; Figure 23).

Many similar studies can be cited, but it is sufficient to stress that the use of generic predominance facies to analyze fossil benthic foraminiferal assemblages is a relatively quick and efficient method that yields highly meaningful interpretations of general paleoenvironments and geologic history through the Cenozoic and Mesozoic Eras.

CONCLUSIONS

On the basis of current knowledge, it is quite clear that the modern distribution of benthic foraminifera, when analyzed at any taxonomic level, is affected by the total physicochemical environment and its evolutionary history. Although not all ecologic factors exert equal influence, and, in some instances a single factor seems to dominate, it is futile to try to reduce all of the patterns of response (for example, the distri-

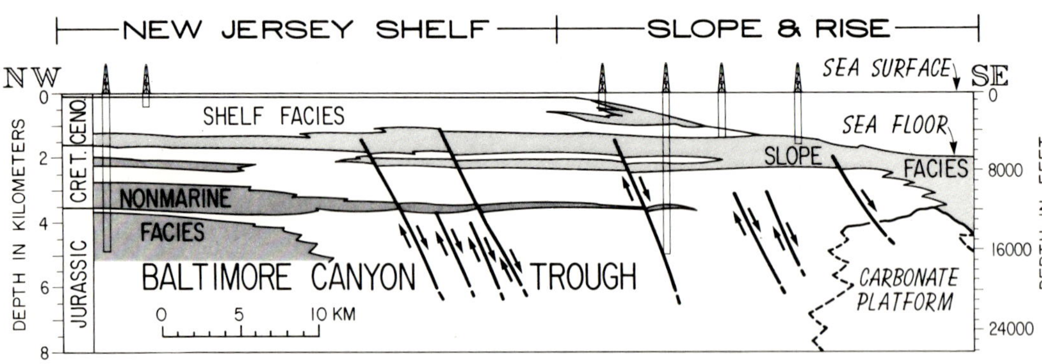

Figure 23
Reconstruction of paleoenvironments and depositional cycles on a section through the Baltimore Canyon Trough, off New Jersey. Paleoenvironments interpreted chiefly from generic predominance facies among the fossil foraminiferal assemblages (data from Poag, 1980).

bution of generic predominance facies) to simple linear relationships. Nevertheless, it is also clear that, especially in the Gulf Coast region, the generic content and generic predominance patterns of modern foraminiferal assemblages can provide much insight into their relationships with physicochemical properties of the environment. Of equal importance, because general species (and generic) diversity gradients have been maintained throughout the Phanerozoic eon (Bambach, 1977), the modern generic predominance patterns serve as useful models for the reconstruction of past environments.

In addition to the physicochemical aspects of the environment (and perhaps overshadowing them in abyssal biotopes), there are enigmatic biological interactions. Bernstein et al. (1978) suggested that biological interactions may provide the best means of understanding dispersion and diversity patterns in the abyss (see also Grassle and Sanders, 1973). Benthic foraminifera collected in box cores from the North Pacific Ocean are as large or larger than individuals of all other metazoan taxa recovered, and the foraminifera may have a more significant effect on the structure and dynamics of abyssal communities than have any of the abyssal microfauna usually studied (Bernstein et al., 1978). We are making advances with regard to these biotic interrelationships among modern benthos, but we may never be able to assess these aspects accurately among fossil benthic assemblages.

Part II

Description and Illustration of Diagnostic Species

TAXONOMY AND DISTRIBUTION

This section provides brief taxonomic notes and a summary of distribution for 138 of the abundant or ecologically diagnostic species belonging to the predominant genera in the Gulf. Each species is illustrated by scanning electron photomicrographs in the following section. The genetic categories into which the species are placed are largely those defined by Loeblich and Tappan (1964), with my modifications (see appropriate notations in text). The synonymies of names applied to each species are not intended to be complete. The original reference is cited along with the most significant name changes and at least one previous citation from the Gulf of Mexico. This atlas is not the proper forum for solving intricate taxonomic problems, but I have pointed out where some of the major problems lie.

Genus *ADERCOTRYMA* Loeblich and Tappan, 1952

ADERCOTRYMA GLOMERATA (H. B. Brady)

Lituola glomerata H. B. Brady, 1878, Ann. Mag. Nat. Hist., ser. 5, vol. 1, p. 433, pl. 20, fig. 1a-c.
Adercotryma glomeratum (H. B. Brady), Parker, 1954, Mus. Comp. Zool., Harvard, Bull., vol. 111, no. 10, p. 486, pl. 1, fig. 18.

Plate 15, Figure 2. Umbilical view showing subquadrate outline, coarse terrigenous and biogenic mineral grains set in finer grained matrix and covered by organic film, and fine-grained umbilical flap of last chamber covering aperture. USNM 304427

Plate 16, Figure 2a. Edge view showing subtriangular outline, coarsest grains on spiral side of test (bottom), and small arched aperture at edge of umbilicus (umbilical flap broken off). USNM 304428

Plate 16, Figure 2b. Spiral view showing coarser particles than in umbilical view, central depression, and lack of organic film covering particles of wall. USNM 304429

Distribution: Scattered component of bathyal and abyssal assemblages; common in agglutinant faunas of *Glomospira* predominance facies; only species of *Adercotryma* known in Gulf. Illustrated specimens from Sigsbee Plain.

Genus *ALVEOLOPHRAGMIUM* Shchedrina, 1936

ALVEOLOPHRAGMIUM sp.

Plate 15, Figure 1. Lateral view showing subcircular outline, planispiral coiling, indistinct sutures, rough surface, and areal aperture at base of final chamber; aperture with upturned ends and slight fine-grained lip surrounding entire margin; wall composed of poorly sorted terrigenous mineral particles; chamber interiors labyrinthic. USNM 304430

Plate 16, Figure 1a. Edge view. USNM 304430

Plate 16, Figure 1b. Close-up view of end of aperture showing fine-grained lip, an extension of inner lining of aperture. USNM 304430

Distribution: Scattered in low numbers in slope and abyssal plain faunas, especially in agglutinant assemblages of *Glomospira* predominance facies. Illustrated specimen from Texas-Louisiana Slope.

Genus *AMMOBACULITES* Cushman, 1910

AMMOBACULITES FILIFORMIS Earland

Haplophragmium agglutinans (d'Orbigny), H. B. Brady, 1884, Rept. Voy. *Challenger*, Zool., vol. 9, p. 301, pl. 32, fig. 23 (21?).
Ammobaculites agglutinans d'Orbigny var. *filiformis* Earland, 1934, *Discovery* Repts., vol. 10, p. 92.
Ammobaculites sp. B Parker, 1954, Mus. Comp. Zool., Harvard, Bull., vol. 111, no. 10, p. 489, pl. 2, fig. 6.

Plate 7, Figure 3. Lateral view showing tight early coil, uniserial adult portion with nearly straight parallel sides, slightly depressed straight sutures, apical aperture with short neck, smoothly polished wall formed of very fine terrigenous mineral particles and covered with thin organic film, and chambers overlapping somewhat at sutures. USNM 304431

Plate 8, Figure 3a. Apertural view showing nearly circular outline of final chamber and small subcircular aperture. Wall roughened around apertural neck where organic layer is missing. USNM 304432

Plate 8, Figure 3b. View from coiled end of test showing arrangement of early chambers, slightly depressed sutures, and smooth, fine-textured wall. USNM 304433

Distribution: Prominent in bathyal and abyssal agglutinant assemblages of *Glomospira* predominance facies. Illustrated specimens from Sigsbee Plain.

AMMOBACULITES sp. cf. *A. FOLIACEUS* (Brady)

cf. *Haplophragmium foliaceum* Brady, 1881, Quart. Jour. Micr. Sci., vol. 21, p. 20 (50).
Ammobaculites cf. *foliaceus* (Brady), Phleger and Parker, 1951, Geol. Soc. America, Mem. 46, pt. II, p. 3, pl. 1, figs. 12, 13.

Plate 49, Figure 4. Lateral view showing early planispiral coil, four chambers in rectilinear part, depressed sutures nearly at right angles to periphery, fine-grained neck of terminal aperture, and chamber walls composed of poorly sorted coarser terrigenous mineral grains set in finer grained matrix. USNM 304434

Plate 50, Figure 4a. Edge view showing strongly compressed test, contrast between fine and coarse particles of test wall, and terminal aperture surrounded by fine-grained neck. USNM 304435

Plate 50, Figure 4b. Apertural view showing lateral compression of test, slitlike aperture, fine-grained construction of neck, and coarser particles in wall. USNM 304436

Plate 50, Figure 4c. Enlargement of aperture showing fine-grained composition of apertural neck; note coccolith at top center for scale. USNM 304436

Distribution: One of most abundant species in *Saccammina-Ammobaculites* predominance facies; scattered additional occurrences in low numbers on inner and middle shelf. Illustrated specimens from Texas Shelf.

AMMOBACULITES sp.

Plate 49, Figure 3. Lateral view showing small early coil, predominant uniserial portion, depressed sutures, apical aperture with distinct neck, and outer wall of coarser, subrounded, poorly sorted terrigenous mineral grains set loosely in finer grained matrix. USNM 304437

Plate 50, Figure 3a. Apertural view showing textural contrast between fine-grained apertural neck and coarser particles in rest of test wall. Note irregular outline of test caused by large mineral grains. USNM 304437

Plate 50, Figure 3b. Enlargement of slitlike aperture and surrounding fine-grained walls of apertural neck. USNM 304437

Distribution: Common in shallow-water marine assemblages on siliciclastic substrates, especially in *Saccammina-Ammobaculites* predominance facies. Illustrated specimen from Texas Shelf.

Genus *AMMODISCUS* Reuss, 1862

AMMODISCUS REESI LeRoy

Ammodiscus reesi LeRoy, 1975, Micropaleont., vol. 21, p. 425, pl. 2, fig. 7.

Plate 7, Figure 1. Lateral view showing nearly circular outline, spiral suture separating successive whorls of single crenulated planispiral chamber, and coarser terrigenous mineral grains interrupting otherwise smooth surface along adumbilical margin of each whorl. USNM 304438

Plate 8, Figure 1a. Edge view showing subrounded periphery, parallel sides, crenulations in final whorl of chamber and circular terminal chamber opening (aperture partly broken); compare organic-film-covered outer surface with rough, filmless area around chamber opening. USNM 304438

Plate 8, Figure 1b. Enlargement of broken chamber at periphery of test contrasting smooth organic film that veneers outer and inner surfaces with coarse-grained middle part of wall. USNM 304438

Distribution: Scattered occurrences in low numbers in bathyal and abyssal assemblages, especially in agglutinant assemblages of *Glomospira* predominance facies. Illustrated specimen from Texas Slope.

AMMODISCUS TENUIS Brady

Ammodiscus tenuis Brady, 1884, Rept. Voy. *Challenger*, Zool., vol. 9, p. 78, pl. 38, figs. 5, 6; Bock, 1971, Miami Geol. Soc., Mem. 1, p. 4, pl. 1, fig. 6.

Plate 7, Figure 2. Lateral view showing smooth, subcircular outline and planispiral coil of single hollow chamber, slightly depressed sutures, wall of fine-grained matrix and scattered larger flat-sided terrigenous mineral grains, and outer surface covered with thin organic film through which larger mineral grains protrude; where chamber wall is broken, coarser construction of wall is evident. USNM 304439

Plate 8, Figure 2a. Edge view showing flattened test, arched chamber opening, rounded periphery, and contrast between organic surface film and coarser grained terminal wall (aperture broken). USNM 304439

Plate 8, Figure 2b. Enlargement of outer wall surface showing organic film surrounding mineral grains (organic film missing in patch at lower left). USNM 304439

Distribution: Low numbers in agglutinant assemblages of *Glomospira* predominance facies typical of submarine prominences in bathyal and abyssal depths. Illustrated specimen from East Mexican Slope.

Genus *AMMONIA* Brünnich, 1772

AMMONIA PARKINSONIANA (d'Orbigny) forma *TEPIDA*

Rotalia beccarii (Linné) var. *tepida* Cushman, 1926, Carnegie Inst. Washington, Publ. 344, p. 79, pl. 1.

"*Rotalia*" *beccarii* (Linné) var. *tepida* Cushman, Phleger and Parker, 1951, Geol. Soc. America, Mem. 46, pt. II, p. 23, pl. 12, figs. 7a, b.

Ammonia parkinsoniana (d'Orbigny) forma *tepida*, Poag, 1978, Gulf Coast Assoc. Geol. Socs., Trans., vol. 28, p. 397, pl. 1, figs. 1-4, 10-12, 17, 18.

Plate 45, Figure 2. Umbilical view showing subcircular lobate outline, seven chambers of final whorl, deep wide umbilicus, pustulose inner ends of chambers, densely perforate chamber walls, imperforate surfaces of final chamber along depressed sutures, and lipped openings of preceding chambers near umbilicus. USNM 304440

Plate 46, Figure 2a. Spiral view showing subcircular lobate outline, depressed spiral suture, raised limbate early intercameral sutures, depressed later intercameral sutures, densely perforate walls, and apparent openings into chambers near junction of spiral and intercameral sutures of last four chambers. USNM 304441

Plate 45, Figure 2b. Edge view showing biconvex outline, depressed sutures of spiral side, and pustulose umbilical region. USNM 304442

Distribution: Forma *tepida* is small, thin-walled ecophenotype of species *A. parkinsoniana*; its morphologic expression depends closely upon variations in environmental parameters, such as salinity; copredominant in *Ammonia* and *Ammonia-Elphidium* predominance facies, in biotopes characterized by optimum environmental conditions that accelerate reproductive maturity (Poag, 1978). Illustrated specimens from San Antonio Bay, Texas.

AMMONIA PARKINSONIANA (d'Orbigny) forma *TYPICA*

Rosalina parkinsoniana d'Orbigny 1939, Foram. de Cuba, vol. 8, p. 99, pl. 4, figs. 25-27.

"*Rotalia*" *beccarii* (Linné) var. *parkinsoniana* (d'Orbigny), Phleger and Parker, 1951, Geol. Soc. America, Mem. 46, pt. II, p. 23, pl. 12, figs. 6a, b.

Ammonia parkinsoniana (d'Orbigny) forma *typica*, Poag, 1978, Gulf Coast Assoc. Geol. Socs., Trans., vol. 28, p. 397, pl. 1, figs. 5-9, 13-16, 19-21.

Plate 45, Figure 1. Umbilical view showing characteristic thick plug in wide, deep umbilicus, imperforate spinose umbilical apertural flaps, sutures deeply depressed near umbilicus but limbate and raised toward periphery, imperforate apertural shoulder of final chamber, and densely perforate chamber walls. USNM 304443

Plate 46, Figure 1a. Spiral view showing tight coiling, subcircular lobate outline, thick imperforate raised sutures except in final two chambers, tiny aberrant subcircular apertures along spiral suture of final chamber, and calcite deposition thickest on earliest sutures. USNM 304444

Plate 46, Figure 1b. Edge view showing umbilical spines on left side of test, thick imperforate raised sutures at right, and periphery totally perforate except at sutures. USNM 304445

Distribution: Forma *typica* is larger, thicker walled ecophenotype of species *A. parkinsoniana*; its morphologic expression depends closely on variations in environmental parameters such as salinity; copredominant in *Ammonia* and *Ammonia-Elphidium* predominance facies in biotopes where environmental conditions are minimal and reproductive maturity is delayed (Poag, 1978). Illustrated specimens from San Antonio Bay, Texas.

AMMONIA PAUCILOCULATA (Phleger and Parker)

"*Rotalia*" *pauciloculata* Phleger and Parker, 1951, Geol. Soc. America, Mem. 46, pt. II, p. 23, pl. 12, figs. 8a, b; 9a, b.

Plate 45, Figure 3. Umbilical view showing subcircular lobate outline, six chambers in final whorl, depressed sutures, intricate system of spinose pustules and ridges in narrow umbilicus, apertural openings beneath broken umbilical extension of final chamber, imperforate areas along sutures, and openings of previous secondary apertures along sutures toward umbilicus. USNM 304446

Plate 46, Figure 3a. Spiral view showing subcircular lobate outline, wide imperforate areas along sutures, and distinct subperipheral depression concentric to carinate periphery. USNM 304447

Plate 46, Figure 3b. Edge view showing inflated biconvex outline, rounded, partly perforate carina along periphery, wide imperforate areas surrounding sutures on spiral side, spinose tubercles projecting around umbilicus, and part of low apertural slit near periphery at base of final chamber. USNM 304448

Distribution: Rare to common in *Ammonia* and *Ammonia-Elphidium* predominance facies along Texas inner shelf; not reported elsewhere in Gulf. Illustrated specimens from Texas Shelf.

Genus *AMMOTIUM* Loeblich and Tappan, 1953

AMMOTIUM SALSUM (Cushman and Brönnimann) forma *DILATATUM*

Ammobaculites dilatatus, Cushman and Brönniman, 1948, Cushman Lab. Foram. Res., Contr., vol. 24, pt. 2, p. 39, pl. 7, figs. 10, 11; Parker, Phleger and Peirson, 1953, Cushman Found. Foram. Res., Spec. Publ. No. 2, p. 5, pl. 1, fig. 13 (?14, 15).

Ammotium salsum (Cushman and Brönnimann) forma *dilatatum* Cushman and Brönnimann, Poag, 1978, Gulf Coast Assoc. Geol. Socs., Trans., vol. 28, p. 405.

Plate 51, Figure 1. Lateral view showing juvenile coil becoming uniserial as adult, distinct deep sutures, terminal aperture, and rugose exterior walls formed from large tabular terrigenous mineral grains cemented in finer grained matrix. USNM 304449

Plate 52, Figure 1a. Lateral view showing remnant of final chamber, diagonal sutures, rugose exterior wall, and flat surfaces of large tabular mineral grains arranged to face outward. USNM 304450

Plate 52, Figure 1b. Apertural view showing subcircular outline of final chamber, terminal aperture with broken margins, elongate opening of previous aperture, and tabular mineral grains with flat surfaces facing outward. USNM 304451

Distribution: Abundant in estuarine and near-shore marine assemblages, especially of *Ammotium* predominance facies; this ecophenotype appears to prefer coarse-grained substrates (Poag, 1978). Illustrated specimens from San Antonio Bay, Texas.

AMMOTIUM SALSUM (Cushman and Brönnimann) forma *EXILE*

Ammobaculites exilis Cushman and Brönnimann, 1948, Cushman Found. Foram. Res., Contr., vol. 24, pt. 2, p. 39, pl. 7, fig. 9.

Ammobaculites salsus Cushman and Brönnimann, Parker, Phleger, and Peirson, 1953, Cushman Found. Foram. Res., Spec. Publ., No. 2, p. 5, pl. 1, figs. 17-21, 23-25 (not fig. 22).

Ammobaculites exilis Cushman and Brönnimann, Phleger, 1954, Amer. Assoc. Petrol. Geol., Bull., vol. 38, no. 4, p. 635, pl. 1, fig. 6.

Ammotium salsum (Cushman and Brönnimann) forma *exile* Cushman and Brönnimann, Poag, 1978, Gulf Coast Assoc. Geol. Socs., Trans., vol. 28, p. 405, pl. 5, figs. 11-32, 34-39.

Plate 51, Figure 4. Lateral view showing overlapping chambers, indistinct, depressed diagonal sutures, large early coil, and wall of poorly sorted large terrigenous mineral grains in finer grained matrix. USNM 304452

Plate 52, Figure 4a. Lateral view showing early coil, diagonal sutures, apical aperture, wide variety of grain sizes used in test wall, and depression in fine-grained cement near center of coil where larger grain has been dislodged. USNM 304453

Plate 52, Figure 4b. Apertural view showing fine-grained texture of inner test lining that protrudes to form apertural neck; test compressed laterally. USNM 304454

Distribution: Especially abundant in *Ammotium* predominance facies in upper reaches of estuaries; this ecophenotype appears to prefer coarse-grained substrates (Poag, 1978). Illustrated specimens from San Antonio Bay, Texas.

AMMOTIUM SALSUM (Cushman and Brönnimann) forma *FRAGILE*

Ammotium fragile Warren, 1957, Cushman Found. Foram. Res., Contr., vol. 8, pt. 1, p. 32, pl. 3, figs. 14, 15.
Ammotium salsum Cushman and Brönnimann forma *fragile* Warren, Poag, 1978, Gulf Coast Assoc. Geol. Socs., Trans., vol. 28, p. 405.

Plate 51, Figure 2. Lateral view showing broad early coil, and beginning of rectilinear portion, slightly lobate outline, poorly sorted terrigenous mineral grains of outer wall, indistinct diagonal sutures, and elongate aperture at apex of final chamber. USNM 304455

Plate 52, Figure 2a. Lateral view showing subcircular early coil, first chamber of rectilinear portion, and contrast between large mineral grains and finer grained matrix; last chamber broken at junction with coil. USNM 304456

Plate 52, Figure 2b. Apertural view showing broken apertural walls of last chamber, lateral compression of test, rounded periphery, and poorly sorted grains of test wall. USNM 304457

Distribution: Abundant in upper reaches of estuaries, bays, and near river mouths where freshwater influence is significant, especially in *Ammotium* predominance facies (Poag, 1978). Illustrated specimens from San Antonio Bay, Texas.

AMMOTIUM SALSUM (Cushman and Brönnimann) forma *TYPICUM*

Ammobaculites salsum Cushman and Brönnimann, 1948, Cushman Found. Foram. Res., Contr., vol. 24, pt. 1, p. 16, pl. 3, figs. 7-9; Parker, Phleger and Peirson, 1953, Cushman Found. Foram. Res., Spec. Publ. No. 2, p. 5, pl. 1, fig. 22 (not figs. 17-21, 23-25).
Ammotium salsum Cushman and Brönnimann forma *typicum*, Poag, 1978, Gulf Coast Assoc. Geol. Socs., Trans., vol. 28, p. 405, pl. 5, figs. 1-10, 33.

Plate 51, Figure 3. Lateral view showing slender elongate megalospheric test, small, early coil, strongly overlapping chambers in rectilinear part, depressed diagonal sutures, and wall of well-sorted coarse terrigenous mineral particles embedded in finer grained matrix. USNM 304458

Plate 52, Figure 3a. Lateral view showing thin, early portion of microspheric test, overlapping chambers, diagonal sutures, gaping (partly broken) aperture at apex of final chamber, and coarse, poorly sorted mineral particles in wall. USNM 304459

Plate 52, Figure 3b. Apertural view showing lateral compression of final chambers, subcircular apical aperture, finer particles composing apertural lip, and early coil protruding (out of focus) at left. USNM 304460

Distribution: Predominant ecophenotype of *A. salsum* in *Ammotium* predominance facies in upper reaches of estuaries and bays around Gulf, except in and adjacent to Campeche carbonate province; this ecophenotype appears to prefer fine-grained substrates (Poag, 1978). Illustrated specimens from San Antonio Bay, Texas.

Genus *AMPHISTEGINA* d'Orbigny, 1826

AMPHISTEGINA GIBBOSA d'Orbigny

Amphistegina gibbosa d'Orbigny, 1839, Foram. de Cuba, vol. 8, p. 120, pl. 8, figs. 1-3.
Amphistegina lessonii d'Orbigny, Phleger and Parker, 1951, Geol. Soc. America, Mem. 46, pt. II, p. 26, pl. 13, figs. 13a, b, 14a, b; pl. 14, figs. 1a, b.
Amphistegina gibbosa d'Orbigny, Poag and Sweet, 1972, Texas A&M Univ. Oceanogr. Stud., vol. 3, p. 236, figs. 8-12c; Crouch and Poag, 1979, Jour. Foram. Res., vol. 8, no. 2, p. 85-105, pls. 1-4; Larsen, 1979, Rev. Español. Micropaleont., vol. 10, p. 225, pl. 3, figs. 1, 2, 4, 5; pl. 7, fig. 8.

Plate 47, Figure 2. Umbilical view showing subcircular outline, fourteen pairs of chambers in final whorl, densely perforate chamber walls, imperforate umbo and radiating sutures of supplementary chambers, narrow aperture near periphery with tuberculate lip and narrow apertural face, aligned tubercles surrounding aperture, and sharp bend in radiating sutures about halfway between umbo and periphery. USNM 304461

Plate 48, Figure 2a. Spiral view showing nineteen chambers in final whorl, sweeping imperforate sutures that recurve sharply near periphery in dog-leg fashion, and imperforate central boss. USNM 304462

Plate 48, Figure 2b. Edge view showing unequally biconvex inflated test, imperforate peripheral band, alignment of apertural tubercles, narrow lip over narrow aperture, and corrosion caused by boring microorganisms. USNM 304463

Distribution: Predominant in *Amphistegina* predominance facies on submerged banks of northern and eastern Gulf, where it forms chiefly relict assemblages, but also occurs as sparse living populations; also abundant on emergent reefs of southern Gulf in Miliolid-*Archaias-Homotrema* predominance facies; only species of *Amphistegina* known in Gulf. Illustrated specimens from West Flower Garden Bank, Texas Shelf.

Genus *ARCHAIAS* de Montfort, 1808

ARCHAIAS ANGULATUS (Fichtell and Moll)

Nautilus angulatus Fichtell and Moll, 1803, Test. Micro., p. 112, pl. 21.
Orbiculina compressa d'Orbigny, 1839, Foram. de Cuba, vol. 8, p. 73, pl. 8, figs. 4-7.
Archaias compressus (d'Orbigny) Phleger and Parker, 1951, Geol. Soc. America, Mem. 46, pt. II, p. 11, pl. 6, figs. 3-5.
Archaias angulatus (d'Orbigny), Bock, 1971, Miami Geol. Soc., Mem. 1, p. 35, pl. 14, figs. 1-3; Wantland, 1975, Amer. Assoc. Petrol. Geol., Stud. Geol., No. 2, p. 392, figs. 9o-r; 12b; 16j, k, m, t.

Plate 47, Figure 4. Lateral view of adult showing early planispiral coil becoming completely enclosed by annular chambers in adult, and elongate chambers separated into chamberlets visible where outer chamber wall is missing. USNM 304464

Plate 48, Figure 4a. Edge view of adult showing truncate flattened periphery, subparallel rows of rimmed apertures, nearly parallel sides of test, and early coil protruding slightly from either side. USNM 304465

Plate 48, Figure 4b. Lateral view of juvenile specimen showing early planispire, densely perforate chamber walls, imperforate sutures, and chamberlets where peripheral wall is broken. USNM 304466

Plate 48, Figure 4c. Edge view of juvenile showing rounded periphery, flattened apertural face of final chamber that encompasses earlier part of coil, coarse perforations of lateral chamber walls, finer perforations on apertural face, and rows of rimmed apertures. USNM 304467

Distribution: Prominent epiphyte on marine grasses on carbonate shelves, reefs and banks; especially notable in Florida Bay, and on West Florida and Campeche shelves; has a quite variable morphology, and various forms of it have been assigned to several different species and genera; in my view, all are conspecific and *A. angulatus* is only species of *Archaias* in Gulf; predominant form in *Archaias* predominance facies, and one of three predominant species in Miliolid-*Archaias*-*Homotrema* predominance facies of living reefs and carbonate banks; common in Miliolid-Soritid and *Asterigerina*-Soritid predominance facies. Illustrated specimens from central Campeche Shelf.

Genus *ARTICULINA* d'Orbigny, 1826

ARTICULINA PACIFICA Cushman

Articulina sulcata Brady (not Reuss), 1884, Rept. Voy. *Challenger*, Zool., vol. 9, p. 183, pl. 12, figs. 12, 13.
Articulina pacifica Cushman, 1944, Cushman Lab. Foram. Res., Spec. Publ. No. 10, p. 17, pl. 14, figs. 14-18; Bock, 1971, Miami Geol. Soc. Mem. 1, p. 32, pl. 13, fig. 6.

Plate 61, Figure 3. Lateral view showing partly parallel widely spaced costae, flaring lipped aperture at left side, and undulating and anastomosing costae on final chamber. USNM 304468

Plate 62, Figure 3a. Lateral view showing peripherally tangential costae in final and penultimate chambers. USNM 304469

Plate 62, Figure 3b. Edge view showing flaring lipped aperture and tangential costae converging from either side toward sharply angular periphery. USNM 304470

Distribution: Abundant in Miliolid predominance facies and other biofacies from carbonate-rich biotopes. Illustrated specimens from eastern Campeche Shelf.

Genus *ASTERIGERINA* d'Orbigny, 1839

ASTERIGERINA CARINATA d'Orbigny

Asterigerina carinata d'Orbigny, 1839, Foram. de Cuba, vol. 8, p. 118, pl. 5, fig. 23; pl. 6, figs. 1, 2; Bock, 1971, Miami Geol. Soc., Mem. 1, p. 54, pl. 19, fig. 12; pl. 20, fig. 1.

Plate 47, Figure 1. Umbilical view showing nearly circular outline, large imperforate central boss, stellate pattern of supplementary chambers, elongate raised ridges below aperture, imperforate apertural flap with lip, imperforate carina, and large pores at edge of apertural flap. USNM 304471

Plate 48, Figure 1a. Spiral view showing imperforate carina, tangential sutures, and densely perforate chamber walls. USNM 304472

Plate 48, Figure 1b. Edge view showing planoconvex shape, protruding calcite mounds and ridges in and around aperture, imperforate carina and apertural face, and coarsely perforate apertural flap. USNM 304473

Distribution: One of predominant forms on central West Florida Shelf, especially in *Asterigerina*, *Rosalina*-*Asterigerina*, *Asterigerina*-Miliolid, and *Asterigerina*-Soritid predominance facies, and on scattered shallow sandy banks around periphery of Gulf; only species of *Asterigerina* recorded from Gulf. Illustrated specimens from central Campeche Shelf.

Genus *BATHYSIPHON* M. Sars, 1872

BATHYSIPHON sp.

Plate 15, Figure 3. Lateral view showing elongate tubular form (both ends broken), test wall composed primarily of fragments of planktic foraminiferal specimens embedded in fine-grained matrix (several fragments are parts of agglutinant benthic specimens), and tendency for flattened sides of fragments to parallel outer surface of test. USNM 304474

Plate 16, Figure 3a. End view. USNM 304474

Plate 16, Figure 3b. Enlargment of outer wall surface. USNM 304474

Distribution: Scattered occurrences in low numbers on slope and rise, Mississippi Fan, and Sigsbee and Florida plains; occasionally a member of *Glomospira* predominance facies. Illustrated specimen from Louisiana-Texas Slope.

Genus *BIGNERINA* d'Orbigny, 1826

BIGENERINA IRREGULARIS Phleger and Parker

Bigenerina irregularis Phleger and Parker, 1951, Geol. Soc. America, Mem. 46, pt. II, p. 4, pl. 1, figs. 16-21.

Plate 33, Figure 3. Lateral view showing elongate outline of test, flattened biserial juvenile part built at angle to rectilinear adult part, adult chambers increasing in size as added, distinct sutural depressions, and coarse terrigenous mineral grains embedded in finer grained matrix. USNM 304475

Plate 34, Figure 3a. Apertural view showing subcircular outline of test and contrast between fine-grained walls of apertural neck and coarse-grained walls of final chamber. USNM 304476

Plate 34, Figure 3b. Prolocular view showing small biserial juvenile part of test set at angle to later chambers and increase in grain size in adult chambers. USNM 304477

Distribution: Predominant species in *Bigenerina* predominance facies of Texas middle and outer shelf, where it may form a relict accumulation; also common in other shelf facies throughout Gulf, especially on siliciclastic substrates.

BIGENERINA sp. cf. *B. TEXTULARIOIDEA* (Goës)

cf. *Clavulina textularioidea* Goës, 1894, K. Svenska Vetenskapakad. Handl., vol. 25, no. 9, p. 42, pl. 8, figs. 387-399.
Bigenerina textularoidea [sic] (Goës), Bock, 1971, Miami Geol. Soc. Mem. 1, p. 9, pl. 2, fig. 7.

Plate 33, Figure 4. Lateral view showing elongate rectilinear adult portion, small biserial juvenile portion, last five to six adult chambers nearly same size, and coarse, rounded biogenic carbonate particles set in finer grained matrix. USNM 304478

Plate 34, Figure 4a. Apertural enlargement contrasting grain sizes of apertural neck and chamber wall. USNM 304479

Plate 34, Figure 4b. Apertural view contrasting finer grained particles of apertural lip with coarser grained particles of outer chamber wall. USNM 304479

Distribution: Common to abundant in numerous shelf facies, especially on carbonate-rich substrates. Illustrated specimens from central Campeche Shelf.

Genus *BRIZALINA* Costa, 1856

BRIZALINA ALATA (Seguenza)

Vulvulina alata Seguenza, 1862, Atti Accad. Gioenia Sci. Nat., ser. 2, vol. 18, p. 115, pl. 2, figs. 5, 5a.
Bolivina alata (Seguenza), Andersen, 1951, Dept. Conserv., La. Geol. Survey, Bull. 35, p. 92, pl. 20, figs. 9a, b.

Plate 23, Figure 2. Lateral view showing overlapping güttulate chambers, coarsely perforate walls, and imperforate peripheral carinae and apertural lip. USNM 304480

Plate 24, Figure 2a. Edge view showing imperforate carina, apertural lips, and upper surfaces of chambers; note serrate tooth-plate in aperture. USNM 304481

Plate 24, Figure 2b. Apertural view showing lateral compression of test, imperforate carina and apertural lips, serrate tooth-plate in areal aperture, asymmetry of chambers, and imperforate upper part of chamber walls. USNM 304481

Plate 24, Figure 2c. Prolocular view (proloculus broken) showing asymmetry of chambers and gradually increasing size of chambers from proloculus to final chamber. USNM 304482

Distribution: Persistent in *Brizalina* and *Brizalina-Uvigerina* predominance facies occurring in low frequencies chiefly on slope. Illustrated specimens from Texas-Louisiana Slope.

BRIZALINA ALBATROSSI (Cushman)

Bolivina albatrossi Cushman, 1922, U.S. Nat. Mus., Bull. 104, pt. 3, p. 31, pl. 6, fig. 4; Parker, 1954, Mus. Comp. Zool., Harvard, Bull., vol. 111, no. 10, p. 513, pl. 7, fig. 13.

Plate 23, Figure 5. Lateral view showing distinctly sculptured surface of test, widely scattered coarse pores in upper ends of U-shaped channels that are rimmed by narrow ridges, and imperforate upper surface of final two chambers. USNM 304483

Plate 24, Figure 5a. Edge view showing lack of carina and large proloculus of megalospheric form. USNM 304484

Plate 24, Figure 5b. Apertural view showing rounded periphery, smooth lipped areal aperture with serrate tooth-plate, and imperforate upper surface of final chambers. USNM 304485

Plate 24, Figure 5c. Prolocular view showing sculptured surface and slight twisting of chambers around longitudinal axis. USNM 304486

Distribution: Predominant species of *Brizalina* in deepest part of *Brizalina* predominance facies that occupies Louisiana and eastern Texas Slope; ranges elsewhere in moderate numbers from middle shelf to lower slope with scattered occurrences on Sigsbee Plain; Pflum and Frerichs (1976) reported megalospheric specimens more common on lower slope than microspheric specimens. Illustrated specimens from Campeche Slope.

BRIZALINA BARBATA (Phleger and Parker)

Bolivina barbata Phleger and Parker, 1951, Geol. Soc. America, Mem. 46, pt. II, p. 13, pl. 6, figs. 12a, b, 13.

Plate 23, Figure 1. Lateral view showing wide perforate sutural bands separating coarsely and densely perforate chamber walls, and imperforate carina terminating in elongate curved spine at lateral extremity of each chamber. USNM 304487

Plate 24, Figure 1a. Edge view showing alternate perforate and imperforate bands on chambers, imperforate carinae terminating in spines, and slight twisting of test around longitudinal axis. USNM 304488

Plate 24, Figure 1b. Apertural view showing carinae joining apertural lips, serrate tooth-plate, and imperforate upper surface of last two chambers. USNM 304488

Plate 24, Figure 1c. Prolocular view of megalospheric specimen showing large spherical proloculus and peripheral spines of chambers. USNM 304489

Distribution: Widespread from inner shelf to lower slope but most abundant near shelf edge in *Brizalina* and *Brizalina-Uvigerina* predominance facies. Illustrated specimens from Texas-Louisiana Shelf edge.

BRIZALINA FRAGILIS (Phleger and Parker)

Bolivina fragilis Phleger and Parker, 1951, Geol. Soc. America, Mem. 46, pt. II, p. 13, pl. 6, figs. 14, 23, 24a, b.

Plate 25, Figure 4. Lateral view showing biserial chamber arrangement, elongate narrow outline, depressed imperforate sutures, imperforate carina, four low costae in early third of test, short basal spine, and tendency to develop faint longitudinal ridges between rows of pores. USNM 304490

Plate 26, Figure 4a. Edge view showing compressed shape, imperforate sutures separating perforate chambers, imperforate peripheral carina, basal costae and spine, and slight twist of test around longitudinal axis. USNM 304490

Plate 26, Figure 4b. Apertural view showing imperforate walls around loop-shaped aperture, apertural lip, and serrate tooth-plate. USNM 304491

Plate 26, Figure 4c. Prolocular view showing compressed shape, imperforate carinae, and basal costae radiating as curved ridges from basal spine. USNM 304492

Distribution: Predominant species of *Brizalina* in that part of *Brizalina* predominance facies that occupies shelf edge off western Louisiana; ranges elsewhere in fewer numbers from inner shelf to middle slope. Illustrated specimens from Texas-Louisiana Shelf edge.

BRIZALINA GOËSII (Cushman)

Bolivina goësii Cushman, 1922, U.S. Nat. Mus., Bull. 104, pt. 3, p. 34, pl. 6, fig. 5; Parker, 1954, Mus. Comp. Zool., Harvard, Bull., vol. 111, no. 10, p. 514, pl. 7, fig. 16.

Plate 23, Figure 4. Lateral view showing lenticular outline, broad depressions in chamber walls surrounded by thickened cuspate ridges along sutures, coarse perforations aligned subparallel to intercameral sutures, chambers arranged in chevron pattern forming narrow, acute angles with long axis of test, many pores with raised rims, narrow scalloped ridge along periphery subconcentric with peripheral carina, and fine pits barely discernible among coarser pores on final chamber but obscured by secondary calcification on previous chambers. USNM 304493

Plate 24, Figure 4a. Peripheral view of megalospheric specimen showing bulbous proloculus, narrow peripheral carina, nonperforate pitted apertural lips, and pitting on early chambers among rimmed coarse pores. USNM 304494

Plate 24, Figure 4b. Apertural view showing wide pitted lips surrounding areal aperture and connected to narrow peripheral carina, serrate tooth-plate in aperture, and fine pitting among rimmed coarse pores. USNM 304495

Plate 24, Figure 4c. Prolocular view showing ridging on large proloculus, narrow peripheral carina, scalloped ridge following each side of carina, and fine pitting among coarser pores. USNM 304496

Distribution: Predominant species of *Brizalina* in that part of *Brizalina* predominance facies that occupies Shelf edge off southern Florida; common in other parts of this facies; occurs elsewhere less abundantly from inner shelf to middle slope. Illustrated specimens from Texas-Louisiana Shelf edge.

BRIZALINA LANCEOLATA (Parker)

Bolivina lanceolata Parker, 1954, Mus. Comp. Zool., Harvard, Bull., vol. 111, no. 10, p. 514, pl. 7, figs. 17-20.

Plate 25, Figure 5. Lateral view showing narrow elongate test, several subdued longitudinal ridges in early portion becoming faint in later chambers, elongate pores aligned in depressions between ridges, wide imperforate sutures between early chambers becoming narrower between later chambers, and short spine base on proloculus. USNM 304497

Plate 26, Figure 5a. Edge view showing alternation of perforate and imperforate bands, ridging in lower half of test, and apertural wall broken away on right side of aperture. USNM 304498

Plate 26, Figure 5b. Apertural view showing lateral compression of test, thick lips surrounding interiomarginal aperture, serrate toothplate, and imperforate upper surface of final two chambers. USNM 304499

Plate 26, Figure 5c. Prolocular view showing imperforate proloculus, ridges of early chambers, and rounded, faintly carinate early chambers. USNM 304500

Distribution: One of least common species of *Brizalina* in *Brizalina* and *Brizalina-Uvigerina* predominance facies and in other outer shelf and slope facies; rare on Mississippi Fan and Sigsbee Plain. Illustrated specimens from Texas-Louisian Shelf edge.

BRIZALINA LOWMANI (Phleger and Parker)

Bolivina lowmani Phleger and Parker, 1951, Geol. Soc. America, Mem. 46, pt. II, p. 13, pl. 6, figs. 20a, b; 21.

Plate 25, Figure 3. Lateral view showing somewhat truncate apertural end with protruding apertural lip, subquadrate adult chambers with scattered coarse pores, heavily calcified juvenile chambers, and distinct constriction separating juvenile portion from adult portion of test. USNM 304501

Plate 26, Figure 3a. Edge view showing wedge-shaped outline, coarse scattered pores, noncarinate periphery, and distinct constriction separating juvenile portion from adult portion of test. USNM 304502

Plate 26, Figure 3b. Apertural view showing imperforate protruding apertural lip and serrate toothplate. USNM 304503

Plate 26, Figure 3c. Enlargement of surface of adult chambers showing sieve plates in pores. USNM 304504

Distribution: Abundant in *Brizalina* and *Brizalina-Uvigerina* predominance facies, especially on Texas and Campeche outer shelf and upper slope, where it is predominant species of genus; occurs elsewhere in moderate numbers from inner shelf (and in bays and estuaries) to Sigsbee Plain; large numbers recorded in plankton tows suggest that pelagic transport may account for its ubiquity (Leavesley et al., 1978); distinct constriction separates adult and juvenile portions of test and may signify ontogenetic change from pelagic to benthic habit. Illustrated specimens from Texas-Louisiana Shelf.

BRIZALINA MINIMA (Phleger and Parker)

Bolivina minima Phleger and Parker, 1951, Geol. Soc. America, Mem. 46, pt. II, p. 14, pl. 6, figs. 22a, b; 25; pl. 7, figs. 1, 2.

Plate 23, Figure 3. Lateral view showing heavily calcified test, subparallel sides, distinct longitudinal ridging in early chambers becoming barely detectable in later chambers, basal edge of each chamber containing serrate margin that overlaps and fuses with previous chambers to extend ridges across depressed sutures, peripheral margin of each chamber forming single curved spine, elongate pores concentrated along basal margins of chambers in early half of test, and pores more circular and more broadly scattered over surface of later chambers. USNM 304505

Plate 24, Figure 3a. Edge view of megalospheric specimen showing overlapping and fusing ridges arising along base of early chambers and crossing intercameral sutures. USNM 304506

Plate 24, Figure 3b. Apertural view showing lip along right side of interiomarginal aperture joining serrate tooth-plate, broad faint carina, and imperforate upper surface of final chambers. USNM 304507

Plate 24, Figure 3c. Prolocular view showing large megalospheric proloculus and peripheral spines. USNM 304508

Distribution: Common in *Brizalina* and *Brizalina-Uvigerina* predominance facies on outer shelf and upper slope; elsewhere ranges in low numbers from inner shelf to Sigsbee Plain. Illustrated specimens from Texas-Louisiana Shelf edge.

BRIZALINA ORDINARIA (Phleger and Parker)

Bolivina simplex Phleger and Parker, 1951 (not *B. interjuncta* Cushman var. *simplex* Cushman and Renz, 1941), Geol. Soc. America, Mem. 46, pt. II, p. 14, pl. 7, figs. 4-6.
Bolivina ordinaria Phleger and Parker, 1952, Cushman Found. Foram. Res., Contr., vol. 3, pt. 1, p. 14.

Plate 23, Figure 6. Lateral view showing scalloped lateral margins formed by narrow chambers, two to three rows of large rimmed pores at basal margins of chambers, imperforate upper chamber margins, and depressed sutures. USNM 304509

Plate 24, Figure 6a. Edge view showing depressed sutures, large pores opening along basal margins of chambers, and weakly developed carina. USNM 304510

Plate 24, Figure 6b. Apertural view showing indistinct apertural lip joining curved serrate tooth-plate, and imperforate upper surface of last chambers. USNM 304511

Plate 24, Figure 6c. Prolocular view showing compressed test, imperforate proloculus, large pore openings, and narrowly rounded margins with tendency to form faint carina. USNM 304512

Distribution: Predominant species of genus in that part of *Brizalina* predominance facies that occupies middle part of slope on either side of Mississippi Delta, and off eastern Louisiana, Mississippi, Alabama, and northern Florida; rare elsewhere from inner shelf to lower slope. Illustrated specimens from Texas-Louisiana Shelf edge.

BRIZALINA STRIATULA (Cushman) forma *SPINATA*

Bolivina striatula Cushman var. *spinata* Cushman, 1936, Cushman Lab. Foram. Res., Spec. Publ. No. 6, p. 59, pl. 8, figs. 9a, b.
Bolivina striatula spinata Cushman, Parker, 1954, Mus. Comp. Zool., Harvard, Bull., vol. 111, no. 10, p. 516, pl. 7, fig. 29.

Plate 25, Figure 6. Lateral view showing narrow elongate biserial test, numerous (∿16) high, narrow longitudinal costae separating longitudinal rows of pores, pores concentrated along basal margins of early chambers, pores more widely scattered over later chambers, basal spine, and branching costae on early chambers. USNM 304513

Plate 26, Figure 6a. Edge view showing high, narrow costae and prolocular spine. USNM 304514

Plate 26, Figure 6b. Apertural view showing lenticular outline, apertural lip, and imperforate upper surface of final two chambers. USNM 304515

Plate 26, Figure 6c. Prolocular view showing costate spinose proloculus, high narrow costae running up test, and rounded periphery. USNM 304516

Distribution: Common to abundant in *Brizalina* and *Brizalina-Uvigerina* predominance facies, especially on outer Louisiana Shelf, where it is predominant species of genus; elsewhere occurs in low numbers from inner shelf (including bays and estuaries) to middle slope. Illustrated specimens from Texas Shelf.

BRIZALINA SUBAENARIENSIS MEXICANA (Cushman)

Bolivina subaenariensis Cushman var. *mexicana* Cushman, 1922, U.S. Nat. Mus., Bull., vol. 104, pt. 3, p. 47, pl. 8, fig. 1.
Bolivina subaenariensis mexicana Cushman, Parker, 1954, Mus. Comp. Zool., Harvard, Bull., vol. 111, no. 10, p. 517, pl. 7, fig. 33.

Plate 25, Figure 1. Lateral view showing lenticular outline, four prominent longitudinal costae, prolocular spine, and patches of widely spaced small pores separated by wide imperforate bands. USNM 304517

Plate 26, Figure 1a. Edge view showing prominent longitudinal costae and carina. USNM 304518

Plate 26, Figure 1b. Apertural view showing compressed test, thickened apertural lips, serrate tooth-plate, longitudinal costae, carinate periphery, and perforate final two chambers. USNM 304519

Plate 26, Figure 1c. Prolocular view showing basal spine, eight prominent longitudinal costae, and sharp peripheral carina. USNM 304520

Distribution: One of most abundant and widespread members of *Brizalina* and *Brizalina-Uvigerina* predominance facies; predominant species of genus along shelf edge and upper slope off northern Mexico (excluding Campeche Shelf), and in nearly continuous band from eastern Texas to northern Florida (this band interrupted off Mississippi Delta); low numbers elsewhere from inner shelf to Sigsbee Plain. Illustrated specimens from Texas-Louisiana Shelf edge.

BRIZALINA TRANSLUCENS (Phleger and Parker)

Bolivina translucens Phleger and Parker, 1951, Geol. Soc. America, Mem. 46, pt. II, p. 15, pl. 7, figs. 13, 14a, b.

Plate 25, Figure 2. Lateral view showing narrow, elongate, microspheric test, scalloped margin formed by subquadrate biserial chambers, large pores restricted to narrow band along base of chambers, and faint ridge perpendicular to long axis of test in lower half of early chambers. USNM 304521

Plate 26, Figure 2a. Edge view showing alternating perforate and imperforate bands and slightly twisted test. USNM 304521

Plate 26, Figure 2b. Apertural view showing ovate outline, apertural lip, serrate tooth-plate, and imperforate upper surface of final chambers. USNM 304521

Plate 26, Figure 2c. Prolocular view showing imperforate megalospheric proloculus, coarse perforations along base of chambers, narrow ridges paralleling chamber margins, and slight twisting of test around longitudinal axis. USNM 304521

Distribution: One of least abundant members of *Brizalina* and *Brizalina-Uvigerina* predominance facies; ranges from inner shelf to Sigsbee Plain. Illustrated specimens from Texas-Louisiana Shelf edge.

Genus *BULIMINA* d'Orbigny, 1826

BULIMINA ACULEATA d'Orbigny

Bulimina aculeata d'Orbigny, 1826, Ann. Sci. Nat., vol. 7, p. 269, no. 7; Parker, 1954, Mus. Comp. Zool., Harvard, Bull., vol. 111, no. 10, p. 510, pl. 6, fig. 19.

Plate 21, Figure 1. Lateral view showing inflated chambers, dense, fine perforations, numerous large spines covering early chambers and base of final chambers, and prominent flexed, serrate tooth-plate. USNM 304522

Plate 22, Figure 1a. Apertural view showing triserial chamber arrangement and prominent flexed, serrate tooth-plate in apertural depression. USNM 304523

Plate 22, Figure 1b. Prolocular view showing subcircular outline of test and subtriangular cross sections of prominent spines. USNM 304524

Distribution: Abundant in *Bulimina* and *Bulimina-Brizalina* predominance facies of middle and lower slope, where it is copredominant with *B. alazanensis*; occurs elsewhere in low numbers from inner shelf to Sigsbee Plain. Illustrated specimens from Campeche Slope.

BULIMINA ALAZANENSIS Cushman

Bulimina alazanensis Cushman, 1927, Jour. Paleont., vol. 1, p. 161, pl. 25, fig. 4; Parker, 1954, Mus. Comp. Zool., Harvard, Bull., vol. 111, no. 10, p. 510, pl. 6, fig. 21.

Plate 21, Figure 3. Lateral view showing seven prominent layered longitudinal costae, basal spine, scattered large pores occupying troughs between costae, final layer of third costa from left extending from final chamber margin down three-quarters length of test to where costa is broken (in unbroken specimens, costae of each successive chamber overlap previous costae and extend to initial end of test), outer part of apertural tooth-plate, and faint striations along upper rim of aperture. USNM 304525

Plate 22, Figure 3a. Apertural view showing ten prominent layered costae, faintly serrate inner ends of flexed tooth-plate in apertural depression, and imperforate upper surface of final two chambers. USNM 304526

Plate 22, Figure 3b. Prolocular view showing radiating pattern of thirteen prominent layered costae, and steplike pattern of adjacent chambers. USNM 304527

Distribution: Abundant and widespread in *Bulimina* and *Bulimina-Brizalina* predominance facies of middle and lower slope, where it is copredominant with *B. aculeata*; ranges elsewhere in low numbers from outer shelf to Sigsbee Plain. Illustrated specimens from middle Campeche Shelf edge.

BULIMINA MARGINATA d'Orbigny

Bulimina marginata d'Orbigny, 1826, Ann. Sci. Nat., vol. 7, p. 269, no. 4, pl. 12, figs. 10-12; Parker, 1954, Mus. Comp. Zool., Harvard, Bull., vol. 111, no. 10, p. 510, pl. 6, fig. 20.

Plate 21, Figure 2. Lateral view showing conical outline, inflated chambers whose overhanging margins extend as large elongate spines, and finely perforate walls. USNM 304528

Plate 22, Figure 2a. Apertural view showing serrate tubular tooth-plate in apertural depression. USNM 304529

Plate 22, Figure 2b. Prolocular view showing triserial chamber arrangement, prominent spines extending from overhanging edges of chambers, and triangular cross-section of most spines. USNM 304530

Distribution: Common in *Bulimina* and *Bulimina-Brizalina* predominance facies all along middle slope; tends to be most abundant on outer shelf and upper slope; low numbers reported from inner and middle shelf and lower slope. Illustrated specimens from Texas-Louisiana Shelf edge.

BULIMINA SPICATA Phleger and Parker

Bulimina spicata Phleger and Parker, 1951, Geol. Soc. America, Mem. 46, pt. II, p. 16, pl. 7, figs. 25a-c; 30, 31.

Plate 21, Figure 5. Lateral view showing narrow elongate outline, inflated final chambers, coarsely perforate wall, prominent aligned costae in early half of test that terminate in sharp overhanging spines, serrate tubular tooth-plate in apertural depression, and faint striations on apertural shoulder. USNM 304531

Plate 22, Figure 5a. Apertural view showing roundly triangular outline, imperforate upper surface of final chambers, and broken tubular tooth-plate. USNM 304532

Plate 22, Figure 5b. Prolocular view showing triserial chamber arrangement, perforate wall, overhanging chamber margins, and spinose tips of costae. USNM 304533

Distribution: Sparse in *Bulimina* and *Bulimina-Brizalina* predominance facies, where most abundant; rare elsewhere from shelf to Sigsbee Plain but more persistent on Texas-Louisiana Shelf than other shelf biotopes. Illustrated specimens from middle Campeche Shelf edge.

BULIMINA STRIATA MEXICANA Cushman

Bulimina striata d'Orbigny var. *mexicana* Cushman, 1922, U.S. Nat. Mus., Bull. 104, pt. 3, p. 95, pl. 21, fig. 2.
Bulimina striata mexicana Cushman, Parker, 1954, Mus. Comp. Zool., Harvard, Bull., vol. 111, no. 10, p. 511, pl. 6, fig. 24.

Plate 21, Figure 4. Lateral view showing conical outline, overhanging chambers with each strong longitudinal costa ending in narrow spine, pores in troughs between costae and occasionally on costae, and serrate tubular tooth-plate. USNM 304534

Plate 22, Figure 4a. Apertural view showing subcircular outline, smooth, porous upper surface of final three chambers, and serrate tubular tooth-plate. USNM 304535

Plate 22, Figure 4b. Prolocular view showing spinose ends of costae (many spine tips broken), and pores occupying intervening depressions. USNM 304536

Distribution: Common in *Bulimina* and *Bulimina-Brizalina* predominance facies along middle and lower slope; low numbers on shelf, upper slope, Sigsbee Plain, and Mississippi Fan. Illustrated specimens from East Mexican Slope.

Genus *BULIMINELLA* Cushman, 1911

BULIMINELLA ELEGANTISSIMA (d'Orbigny)

Bulimina elegantissima d'Orbigny, 1829, Voy. Amér. Mérid., vol. 5, pt. 5, "*Foraminifères*", p. 51, pl. 7, figs. 13, 14.
Buliminella elegantissima (d'Orbigny) Phleger and Parker, 1951, Geol. Soc. America, Mem. 46, pt. II, p. 17, pl. 8, figs. 3, 4.

Plate 33, Figure 2. Lateral view showing fusiform shape, densely perforate chambers, imperforate depressed sutures, imperforate proloculus, broad flattened final chamber face becoming imperforate around aperture, and tubular rim of aperture. USNM 304537

Plate 34, Figure 2a. Apertural view showing subcircular outline, flattened apertural face, and deep apertural depression. USNM 304538

Plate 34, Figure 2b. Apertural enlargement showing tubular apertural rim protruding from apertural depression. USNM 304538

Distribution: Less abundant of two Gulf species of *Buliminella*; distribution similar to that of *B. morgani* in western and southern Gulf, where most common on inner and middle Shelf; infrequent in eastern Gulf; reported in low numbers in bays and estuaries and as deep as middle slope. Illustrated specimens from Louisiana Shelf.

BULIMINELLA MORGANI Andersen

Buliminella cf. *B. bassendorfensis* Cushman and Parker, Phleger and Parker, 1951, Geol. Soc. America, Mem. 46, pt. II, p. 17, pl. 8, figs. 1, 2.
Buliminella morgani Andersen, 1961, Dept. Conserv. La. Geol. Survey, Bull. 35, pt. II, p. 87, pl. 19, fig. 10.

Plate 33, Figure 1. Lateral view showing narrow elongate outline, finely perforate wall, narrow imperforate band at upper margin of each chamber, prominent proloculus, and stubby tooth in aperture. USNM 304539

Plate 34, Figure 1a. Apertural view showing subcircular outline, triserial chamber arrangement in adult portion of test, stubby tooth in aperture, and imperforate apertural shoulder. USNM 304540

Plate 34, Figure 1b. Prolocular view showing partly perforate spherical proloculus and quadriserial chamber arrangement in early part of test. USNM 304541

Distribution: Common in shallow shelf biotopes around Gulf, especially near deltas; becomes predominant species in *Buliminella* predominance facies, which is limited to middle shelf off western Louisiana; reported in low numbers from bays and estuaries, and as deep as middle slope in eastern Gulf. Illustrated specimens from Louisiana Shelf.

Genus *CIBICIDES* de Montfort, 1808

CIBICIDES PROTUBERANS Parker

Cibicides protuberans Parker, 1954, Mus. Comp. Zool., Harvard, Bull., vol. 111, no. 10, p. 542, pl. 12, figs. 13, 14, 16.

Plate 3, Figure 4. Spiral view showing distinct spiral coiling with wide, imperforate sutures, fine but densely distributed perforations, flat attachment surface, and numerous chambers; last chamber broken. USNM 304542

Plate 4, Figure 4a. Umbonal view showing clearly defined imperforate umbonal boss, densely perforate remaining surface, and strongly curved depressed sutures; final two chambers broken. USNM 304543

Plate 4, Figure 4b. Edge view showing concavoconvex shape of flattened test, dense perforations, imperforate peripheral carina, and lipped aperture crossing periphery. USNM 304544

Distribution: Shallow-water member of genus; common in *Cibicidoides* predominance facies of West Florida and Campeche shelves; scattered occurrences reported elsewhere on northern shelf and along West Florida Escarpment. Illustrated specimens from Louisiana Shelf.

CIBICIDES RUGOSUS Phleger and Parker

Cibicides rugosa Phleger and Parker, 1951, Geol. Soc. America, Mem. 46, pt. II, p. 31, pl. 17, figs. 5a, b; 6a, b.

Plate 3, Figure 3. Spiral view showing flattened surface, numerous arcuate chambers whose pores occupy broad pentagonal pits, slightly frilled imperforate carina, and lipped aperture. USNM 304545

Plate 4, Figure 3a. Umbonal view showing subcircular outline and strongly rugose protuberances covering all but last two chambers. USNM 304546

Plate 4, Figure 3b. Edge view showing nearly planoconvex outline, strongly rugose umbonal side, flattened apertural face of final chamber, narrow, imperforate carina, and arched lipped aperture crossing periphery. USNM 304547

Distribution: Deep-water species known chiefly from middle and lower slope; rare in *Cibicides* predominance facies. Illustrated specimens from Campeche Slope.

CIBICIDES WUELLERSTORFI (Schwager)

Anomalina wuellerstorfi Schwager, 1866, Novara-Exped., Geol. Theil., vol. 2, p. 258, pl. 7, figs. 105, 107.
Planulina wuellerstorfi (Schwager), Phleger and Parker, 1951, Geol. Soc. America, Mem. 46, pt. II, p. 33, pl. 19, figs. 1a, b; 2a, b; 3a, b (?pl. 18, figs. 11a, b).
Cibicides wuellerstorfi (Schwager), Parker, 1954, Mus. Comp. Zool., Harvard, Bull., vol. 111, no. 10, p. 544, pl. 13, figs. 3, 6.

Plate 3, Figure 1. Spiral view showing nearly smooth periphery, flat surface, numerous arcuate coarsely perforate narrow chambers, thick imperforate sutures and carina, and fringed apertural lip. USNM 304548

Plate 4, Figure 1a. Umbonal view showing strongly curved, raised sutures, coarse perforations in final four chambers, and nearly smooth periphery. USNM 304549

Plate 4, Figure 1b. Edge view showing planoconvex outline with flattened spiral side, thick peripheral carina, flattened apertural face, and ribbed apertural lip crossing periphery. USNM 304550

Distribution: Deep-water species characteristic of lower slope, rise, and abyssal plains; predominant species of *Cibicides* predominance facies on Mississippi Fan; reported in low numbers from upper and middle slope. Illustrated specimens from Campeche Slope.

CIBICIDES sp.

Plate 3, Figure 2. Spiral view showing flat surface, numerous arcuate chambers, sieve plates in coarse pores, imperforate sutures and peripheral margin, flattened apertural face, and spatulate apertural lip. USNM 304551

Plate 4, Figure 2a. Umbonal view showing heavily calcified surface, depressed curved sutures, and prominent umbonal knob. USNM 304552

Plate 4, Figure 2b. Edge view showing high conical outline with flat spiral side, strong imperforate peripheral carina, distinctly flattened apertural face of last chamber, and arched crenulate apertural lip crossing periphery. USNM 304553

Distribution: Undescribed species whose range is not currently known, but appears to be most common (in low numbers) on outer shelf and slope. Illustrated specimens from Louisiana Slope.

Genus *CIBICIDOIDES* Thalmann, 1939

CIBICIDOIDES CORPULENTUS (Phleger and Parker)

Cibicides robustus Phleger and Parker, 1951 (not *C. robustus* Le Calvez, 1949), Geol. Soc. America, Mem. 46, pt. II, p. 31, pl. 17, figs. 1a, b; 2a, b; 3a, b, 4a, b.
Cibicides corpulentus Phleger and Parker, 1952, Cushman Found. Foram. Res., Contr., vol. 3, pt. 1, p. 14.

Plate 31, Figure 1. Spiral view showing very coarse pitted perforations bordered by imperforate sutural bands, and thickly calcified plug in center of spire surrounded by deep trench. USNM 304554

Plate 32, Figure 1a. Umbonal view showing coarse pitted perforations and heavily calcified early chambers. USNM 304555

Plate 32, Figure 1b. Edge view showing asymmetrical biconvex outline, broadly rounded imperforate periphery, imperforate apertural face, and apertural lip crossing periphery. USNM 304556

Distribution: Large species found in moderate numbers at shelf edge in

Cibicidoides predominance facies; scattered occurrences from middle shelf to lower slope. Illustrated specimens from Texas-Louisiana Shelf edge and Campeche Slope.

CIBICIDOIDES "*FLORIDANUS*" forma *BATHYALIS*

Cibicides aff. *floridanus* (Cushman), Phleger and Parker, 1951, Geol. Soc. America, Mem. 46, pt. II, p. 30, pl. 16, figs. 1a, b (not figs. 2a, b; 3a, b; 4a, b); Parker, 1954, Mus. Comp. Zool., Harvard, Bull., vol. 111, no. 10, p. 541, pl. 12, figs. 5, 9.
Cibicidoides floridanus (Cushman), LeRoy and Levinson, 1974 (not Cushman, 1918), Micropaleontology, vol. 20, p. 16, pl. 8, figs. 7-9.

Plate 29, Figure 1. Spiral side showing numerous narrow, coarsely perforate arcuate chambers, wide imperforate raised sutures, thickened carina and central boss, and aperture with frilled lip. USNM 304557

Plate 30, Figure 1a. Umbonal view showing wide imperforate peripheral shelf, imperforate umbo, narrow perforate chambers, and nearly smooth subcircular outline. USNM 304558

Plate 30, Figure 1b. Edge view showing strongly biconvex shape, distinct peripheral shelf formed between sharp carina and umbo, imperforate apertural face of final chamber, and crenulate apertural lip crossing periphery. USNM 304559

Distribution: The group of modern forms recorded in the literature as *C.* aff. *floridanus* and *C. floridanus* is clearly related to, but not conspecific with *C. floridanus* (Cushman), which was described from the upper Miocene Red Bay formation of Florida. Berggren et al. (1976) have recently suggested that the name *C. mediocris* (Finlay) be applied to some of the modern forms. Whatever the preferred taxonomic disposition may be, there are two distinct but intergrading morphologies among these modern Gulf forms that appear to be ecophenotypes of the species I have termed *C. "floridanus."* Forma *bathyalis* is the larger, thicker walled form that is common in slope facies around the Gulf. Illustrated specimens from Louisiana Shelf.

CIBICIDOIDES "FLORIDANUS" forma *SUBLITTORALIS*

Cibicides aff. *floridanus* (Cushman), Phleger and Parker, 1951, Geol. Soc. America, Mem. 46, pt. II, p. 30, pl. 16, fig. 4a, b (not figs. 1a, b; 2a, b; 3a, b).
Cibicides pseudoungerianus (Cushman), Bandy, 1956 (not Cushman, 1922), U.S. Geol. Survey, Prof. Paper 274-G, p. 193, pl. 31, figs. 8a-c.
Cibicides floridanus (Cushman), Andersen, 1961 (not Cushman, 1918), Dept. Conserv. La. Geol. Survey, Bull. 35, pt. 2, p. 123, pl. 28, figs. 6a-c.
Cibicides cicatricosus (Schwager), Bock, 1971, pl. 22, figs. 3, 4 (not *Anomalina cicatricosa* Schwager, 1866).
Cibicides cf. *pseudoungerianus* (Cushman), Pflum and Frerichs, 1976, J. Foram. Res., Spec. Publ. No. 14, pl. 2, fig. 9; pl. 3, figs. 1, 2.

Plate 29, Figure 2. Spiral view showing slightly lobate periphery, tangential sutures, imperforate carina, coarse pitted perforations, serrate apertural lips, rugose central boss, and chambers whose heights are slightly greater than widths. USNM 304575

Plate 30, Figure 2a. Umbonal view showing slightly lobate periphery, smooth imperforate sutures, and carina. USNM 304576

Plate 30, Figure 2b. Edge view showing highly convex rugose spiral side, smooth, slightly convex umbonal side, periphery rimmed by thick sinuous imperforate carina, and apertural lip crossing periphery. USNM 304577

Distribution: Small, thin-walled form appears to be shallow-water ecophenotype of *C. "floridanus"*;

common member of shelf-edge facies, becoming predominant member of genus in *Cibicidoides* predominance facies at edge of Campeche and West Florida shelves (but concentration there may be relict feature of lowered late Pleistocene or early Holocene sea level); apparently restricted to shelf and upper slope. Illustrated specimens from Texas-Louisiana shelf edge.

CIBICIDOIDES IO (Cushman)

Cibicides pseudoungeriana (Cushman) var. *io* Cushman, 1931, U.S. Nat. Mus., Bull. 104, pt. 8, p. 125, pl. 23, fig. 1 (not fig. 2).
Cibicides io Cushman, Parker, 1954, Mus. Comp. Zool., Harvard, Bull., vol. 111, no. 10, p. 542, pl. 12, figs. 6, 7.

Plate 31, Figure 2. Spiral view showing heavily calcified perforate walls, nearly smooth periphery, deep spiral suture, frilled apertural lip, and nearly straight, radiating intercameral sutures. USNM 304560

Plate 32, Figure 2a. Umbonal view showing smooth surface and periphery, slightly depressed umbo, and rather evenly spaced deeply pitted perforations. USNM 304561

Plate 32, Figure 2b. Edge view showing strong lateral compression of test, rounded periphery, somewhat carinate early coil, flattened apertural face of final chamber, and nodular apertural lip crossing periphery. USNM 304562

Distribution: Persistent but sparse in *Cibicidoides* predominance facies; apparently restricted to middle and outer shelf and upper slope. Illustrated specimens from Texas Shelf.

CIBICIDOIDES MOLLIS (Phleger and Parker)

Cibicides mollis Phleger and Parker, 1951, Geol. Soc. America, Mem. 46, pt. II, p. 30, pl. 16, figs. 7a, b; 8a, b; 9a, b.

Plate 29, Figure 4. Spiral view showing coarse perforations with sieve plates, strongly lobate periphery, few chambers of nearly equal height and width, finely pustulose surface, and flaring frilled apertural lips. USNM 304566

Plate 30, Figure 4a. Umbonal view showing depression at central juncture of sutures, perforate outer band on later chambers, and finely pustulose surface. USNM 304567

Plate 30, Figure 4b. Edge view showing thin lenticular outline, imperforate peripheral carina, coarse perforations with sieve plates, frilled apertural lip crossing periphery, and finely pustulose surface. USNM 304568

Distribution: Persistent low numbers in middle and outer shelf facies around northern and eastern Gulf; rare on upper slope. Illustrated specimens from Texas Shelf.

CIBICIDOIDES ROBERTSONIANUS (H. B. Brady)

Truncatulina robertsonianus H. B. Brady, 1881, Quart. Jour. Micro. Sci., vol. 21, p. 65; 1884, Rept. Voy. *Challenger,* Zool., vol. 9, p. 664, pl. 95, figs. 4a-c.
Cibicides robertsonianus (H. B. Brady), Parker, 1954, Mus. Comp. Zool., Harvard, Bull., vol. 111, no. 10, p. 543, pl. 13, figs. 2, 5.

Plate 5, Figure 1. Lateral view showing nearly circular outline, numerous subquadrate coarsely perforate chambers of final whorl, imperforate sutures and periphery, and peripheral aperture bordered above by undulate lip. USNM 304569

Plate 6, Figure 1a. Umbonal view showing totally imperforate test wall, and subcircular outline. USNM 304570

Plate 6, Figure 1b. Edge view showing nearly planoconvex shape, imperforate convex side, perforate spiral side, and peripheral aperture with slightly thickened rim above it. USNM 304571

Distribution: Deep-water form; low to moderate numbers on slope and rise; occasionally on outer shelf (W. D. Bock, written comm., 1979); often predominant form of genus in *Cibicidoides* predominance facies on Campeche Slope. Illustrated specimens from East Mexico Slope.

CIBICIDOIDES sp. cf. *C. UMBONATUS* (Phleger and Parker)

cf. *Cibicides umbonatus* Phleger and Parker, 1951, Geol. Soc. America, Mem. 46, pt. II, p. 31, pl. 17, figs. 7a, b (?8a, b; 9a, b).

Plate 31, Figure 4. Spiral view showing subcircular outline, coarse perforations, and serrate apertural border. USNM 304572

Plate 32, Figure 4a. Umbonal view showing imperforate periphery and rugose umbonal protuberance. USNM 304573

Plate 32, Figure 4b. Edge view showing lenticular outline, rugose umbo, thick imperforate carina, and aperture crossing periphery. USNM 304574

Distribution: Present in low numbers in shelf-edge *Cibicidoides* predomance facies; scattered occurrences on middle shelf and on slope. Illustrated specimens from Texas-Louisiana Shelf edge and slope.

CIBICIDOIDES sp. *A*

Cibicidoides aff. *floridanus* (Cushman), Phleger and Parker, 1951, Geol. Soc. America, Mem. 46, pt. II, p. 30, pl. 16, figs. 2a, b; 3a, b (not figs. 1a, b; 4a, b).

Plate 29, Figure 3. Spiral view showing lobate outline, coarsely perforate chambers whose height is approximately one-half width, imperforate sutures strongly tangential to periphery, and frilled apertural lip. USNM 304578

Plate 30, Figure 3a. Umbonal view showing slightly lobate periphery, imperforate bands on lower edges of later chambers, calcite covering pores of earlier chambers, and umbo depressed slightly below last inflated chamber. USNM 304579

Plate 30, Figure 3b. Edge view showing strong, nearly equal biconvexity, imperforate peripheral carina, fringed apertural lip crossing periphery, and inflated final chamber standing higher than umbo. USNM 304580

Distribution: Previously reported from northwestern Gulf; clearly a species distinct from forms I have termed *C. "floridanus"*; distribution not well known but appears to be limited in moderate numbers to shelf and upper slope. Illustrated specimens from Texas-Louisiana Shelf edge.

CIBICIDOIDES sp. *B*

Plate 31, Figure 3. Spiral view showing slightly lobate periphery, numerous subquadrate coarsely perforate chambers, imperforate peripheral carina, sutures, and umbo, fine pustules around aperture and spiral suture, and granular apertural flap. USNM 304563

Plate 32, Figure 3a. Umbonal view showing nearly smooth periphery and imperforate granular test surface; scattered openings are borings of undetermined origin. USNM 304564

Plate 32, Figure 3b. Edge view showing asymmetrical biconvexity, imperforate granular surface of umbonal side, undulating imperforate carina, and nodular apertural lip crossing periphery. USNM 304565

Distribution: Deep-water form of genus, common on middle and lower slope; ranges onto Sigsbee Plain; rare in shelf-edge *Cibicidoides* predominance facies but is predominant species of genus in some samples in *Cibicidoides* predominance facies on Campeche Slope. Illustrated specimens from Campeche Slope.

Genus *CRIBROSTOMOIDES* Cushman, 1910

CRIBROSTOMOIDES NITIDUM (Goës)

Haplophragmium nitidum Goës, 1896, Mus. Comp. Zool., Harvard, Bull., vol. 29, p. 30, pl. 3, figs. 8, 9.
Alveolophragmium nitidum (Goës), Parker, 1954, ibid, vol. III, no. 10, p. 486, pl. 1, fig. 17.

Plate 9, Figure 2. Lateral view showing planispiral coil, lobate subquadrate outline, deep, open umbilicus, outer chamber walls composed of flattened coarse terrigenous mineral grains set in finer grained matrix and covered by thin organic film, distinct sutures, and remains of wall of broken chamber still visible along side of first chamber of final whorl. USNM 304581

Plate 10, Figure 2a. Edge view showing inflated chambers, lobate biumbilicate outline, distinct sutures, thin slitlike aperture just above base of final chamber surrounded by fine-grained lip, outer wall composed of flattened coarse grains in finer matrix, and outer organic film that obscures sharp edges of mineral grains. USNM 304582

Plate 10, Figure 2b. Enlargement of aperture showing lip composed of fine grains on organic membrane; note that organic film obscures sharp edges of mineral grains on wall below aperture. USNM 304582

Distribution: Scattered occurrences in low numbers on upper slope and deeper biotopes; rare on shelf; typical in low numbers in agglutinant *Glomospira* predominance facies. Illustrated specimens from Louisiana Slope.

CRIBROSTOMOIDES ROTULATUM (H. B. Brady)

Haplophragmium rotulatum H. B. Brady, 1881, Quart. Jour. Micr. Sci., vol. 21, p. 50; 1884 Rept. Voy. *Challenger*, Zool., vol. 9, p. 306, pl. 34, figs. 5, 6.

Plate 11, Figure 3. Lateral view showing planispiral coiling, large, deeply excavated umbilicus, wall constructed of poorly sorted larger terrigenous mineral grains and tests of calcareous foraminifera set in finer grained matrix, last chamber of final whorl about twice diameter of first chamber of final whorl, and fine-grained apertural lips protruding from flattened face of last remaining chamber; final two chambers broken off and only lower apertural lip of each remains along periphery. USNM 304583

Plate 12, Figure 3a. Edge view showing somewhat flattened periphery, deep umbilici, elongate areal aperture, and wall formed largely of tests of planktic foraminifera set in finer matrix. USNM 304584

Plate 12, Figure 3b. Enlargement of aperture showing fine-grained lip composed of coccoliths and coccolith-sized particles. USNM 304584

Distribution: Occasional specimens on slope, rise, Mississippi Fan, and Sigsbee Plain, especially in agglutinant *Glomospira* predominance facies. Illustrated specimens from Louisiana Slope.

CRIBROSTOMOIDES SCITULUM (H. B. Brady)

Haplophragmium scitulum H. B. Brady, 1881, Quart. Jour. Micr. Sci., vol. 21, p. 50; 1884 Rept. Voy. *Challenger*, Zool., vol. 9, p. 308, pl. 34, figs. 11-13.
Alveolophragmium scitulum (H. B. Brady), Parker, 1954, Mus. Comp. Zool., Harvard, Bull., vol. 111, no. 10, p. 487, pl. 1, figs. 20, 21.

Plate 11, Figure 4. Lateral view showing planispiral evolute coiling, deep umbilicus, faint intercameral sutures becoming more distinct between final four chambers, surface composed of large, flattened terrigenous mineral grains in matrix of finer grains, and fine-grained lips of slitlike areal aperture protruding from apertural face of final chamber. USNM 304585

Plate 12, Figure 4a. Edge view showing broadly rounded periphery, flattened apertural face, narrow, sinuous, lipped areal aperture, and depressed umbilici. USNM 304586

Plate 12, Figure 4b. Enlargement of aperture showing wall composed of larger flattened mineral grains in finer matrix covered by thin organic film that obscures sharp angles of mineral grains; apertural lips formed of even finer grained particles. USNM 304586

Distribution: Scattered occurrences in low numbers on slope, rise, Mississippi Fan, and Sigsbee Plain, especially in agglutinant *Glomospira* predominance species. Illustrated specimens from East Mexico Slope.

CRIBROSTOMOIDES SUBGLOBOSUM (G. O. Sars)

Lituola subglobosa M. Sars, 1868 (1869), Vidensk.-Selsk. Christiania, Förhandl., p. 250 (nomen nudum); G. O. Sars, 1872, ibid, p. 253.
Alveolophragmium subglobosum (G. O. Sars), Parker, 1954, Mus. Comp. Zool., Harvard, Bull., vol. 111, no. 10, p. 487, pl. 2, figs. 1, 2.

Plate 11, Figure 2. Lateral view showing subcircular outline, planispiral coil, depressed umbo, outer wall composed of well-sorted larger terrigenous mineral grains set with flat surfaces facing outward in finer grained matrix and covered by thin organic film; outer organic film and most flat-faced mineral grains eroded from final chamber, revealing closely compact grains of middle wall layer. USNM 304587

Plate 12, Figure 2a. Edge view showing ovate biumbilicate outline, outer wall of coarser grains set with flat faces outward in finer matrix and covered with thin organic film, and narrow aperture surrounded by fine-grained lip near base of final chamber; fragments of wall of broken chamber surround apertural region. USNM 304588

Plate 12, Figure 2b. Enlargement of apertural region showing fine-grained composition of apertural lips and outer organic film that obscures sharp edges of large mineral grains in wall below aperture. USNM 304588

Distribution: Scattered occurrences in low numbers on slope, rise, Mississippi Fan, and Sigsbee Plain, especially in agglutinant *Glomospira* predominance facies. Illustrated specimens from East Mexico Slope.

CRIBROSTOMOIDES WIESNERI (Parr)

Labrospira wiesneri Parr, 1950, B.A.N.Z., Antarctic Res. Exped., 1929-1931, Repts., Ser. B, vol. 5, pt. 6, p. 272, pl. 4, figs. 25, 26.
Labrospira arctica Parker, 1952, Mus. Comp. Zool., Harvard, Bull., vol. 106, no. 9, p. 399, pl. 2, figs. 7, 12.
Alveolophragmium wiesneri (Parr), Parker, 1954, ibid., vol. 111, no. 10, p. 488, pl. 1, fig. 23.

Plate 9, Figure 1. Lateral view showing slightly evolute planispiral coil, lobate outline, distinct radiating straight sutures, smooth, thin organic layer covering coarse terrigenous mineral grains (organic layer worn away in parts of final chamber), and everted areal apertural lips protruding from flattened face of final chamber. USNM 304589

Plate 10, Figure 1a. Edge view showing subangular periphery, inflated final chamber, depressed umbilici, lipped areal aperture, and smooth organic film on surface of test. USNM 304590

Plate 10, Figure 1b. Enlargement of aperture showing fine-grained composition of lips; note coarser grains of surrounding chamber wall covered by thin smooth organic film. USNM 304590

Distribution: Persistent low numbers on slope, rise, Mississippi Fan, and Sigsbee Plain, especially in agglutinant *Glomospira* predominance facies. Illustrated specimens from East Mexico Slope.

CRIBROSTOMOIDES sp.

Plate 11, Figure 1. Lateral view showing subcircular outline and poorly sorted coarser terrigenous mineral grains in finer matrix of wall. USNM 304591

Plate 12, Figure 1a. Edge view showing subcircular outline, flattened apertural face containing lipped, lenticular, areal aperture; wall broken beneath aperture. USNM 304592

Plate 12, Figure 1b. Enlargement of aperture showing fine-grained structure of lips; wall broken at lower left of aperture. USNM 304592

Distribution: Scattered occurrences in low numbers on slope, rise, Mississippi Fan, and Sigsbee Plain, especially in agglutinant *Glomospira* predominance facies. Illustrated specimens from East Mexico Slope.

CYCLAMMINA CANCELLATA H. B. Brady

Cyclammina cancellata H. B. Brady, 1876 (M.S.) Proc., Roy. Soc., vol. 25, p. 214 (nomen nudum); 1884, Rept. Voy. *Challenger*, Zool., vol. 9, p. 351, pl. 37, figs. 8-16; Phleger and Parker, 1951, Geol. Soc. America, Mem. 46, pt. II, p. 3, pl. 1, fig. 15.

Plate 9, Figure 3. Lateral view showing subcircular outline, smooth organic film covering coarse terrigenous mineral grains except on apertural face of last chamber and in several eroded spots, and sinuate chambers overlapping at umbilicus. USNM 304593

Plate 10, Figure 3a. Edge view showing subangular periphery and flattened, coarse-textured apertural face above basal apertural slit. USNM 304594

Plate 10, Figure 3b. Enlargement of aperture showing basal apertural slit, wall above composed of coarse, angular mineral grains in finer matrix, wall below containing better sorted smaller grains covered by smooth, thin organic film. USNM 304594

Distribution: Scattered occurrences in low numbers on slope, rise, Mississippi Fan, and Sigsbee Plain; occasionally encountered in agglutinant

Glomospira predominance facies. Illustrated specimens from East Mexico Slope.

Genus *DISCAMMINA* Lacroix, 1932

DISCAMMINA COMPRESSA (Goës)

Lituolina irregularis Roemer var. *compressa* Goës, 1882, K. Svenska Vetenskapakad. Förhandl. Handl., vol. 19, no. 4, p. 141.
Haplophragmoides emaciatum Brady, 1884, Rept. Voy. *Challenger*, Zool., vol. 9, p. 305, pl. 33, figs. 26-28.

Plate 9, Figure 4. Lateral view showing lobate periphery and coarse-grained structure of outer wall. USNM 304595

Plate 10, Figure 4a. Edge view showing severe compression of test, coarse terrigenous mineral grains of wall, and fine-grained lips of areal aperture. USNM 304596

Plate 10, Figure 4b. Enlargement of aperture showing fine-grained apertural lips with coarse-grained chamber walls. USNM 304596

Distribution: Rare occurrences in low numbers on slope, rise, Mississippi Fan, and Sigsbee Plain; occasionally present in agglutinant *Glomospira* predominance facies. Illustrated specimens from East Mexico Slope.

Genus *ELPHIDIUM* de Montfort, 1808

ELPHIDIUM DISCOIDALE (d'Orbigny) forma *TRANSLUCENS*

cf. *Elphidium translucens* Natland, 1933, Scripps Inst. Oceanogr., Tech. Ser., Bull., vol. 4, p. 144, pl. 5, figs. 3, 4.
Elphidium cf. *E. translucens* Natland, Andersen, 1961, Dept. Conserv., La. Geol. Survey, Bull. 35, pt. 2, p. 110, pl. 18, figs. 10a, b.
Elphidium discoidale (d'Orbigny), Bock, 1971, Miami Geol. Soc., Mem. 1, p. 56, pl. 20, figs. 9, 10.

Plate 35, Figure 2. Lateral view showing involute planispiral coil, densely perforate chambers, imperforate retral bands with large fossettes, and imperforate umbo containing canal openings. USNM 304597

Plate 36, Figure 2a. Edge view showing lenticular outline, slightly swollen umbos, densely perforate chambers, sparsely perforate peripheral margin, imperforate apertural face with basal cribrate aperture, and imperforate retral bands containing fossettes. USNM 304598

Plate 36, Figure 2b. Enlargement of apertural area showing lipped multiple apertural openings and intervening clumps of spines. USNM 304598

Distribution: Along with forma *typicum*, the copredominant form of *Elphidium* in *Elphidium* and *Elphidium-Hanzawaia* predominance facies on Texas Shelf, where it may form a relict accumulation; elsewhere, occurs in varying abundance across continental shelf and extends into bays and estuaries; forma *translucens* is thin-walled ecophenotype resulting from accelerated reproductive maturity in favorable environmental conditions (Poag, 1978). Illustrated specimens from Texas Shelf.

ELPHIDIUM DISCOIDALE (d'Orbigny) forma *TYPICUM*

Polystomella discoidalis d'Orbigny, 1839, Foram. de Cuba, vol. 8, p. 56, pl. 6, figs. 23, 24.
Elphidium discoidale (d'Orbigny), Phleger and Parker, 1951, Geol. Soc. America, Mem. 46, pt. II, p. 10, pl. 5, fig. 10 (?fig. 11).
Elphidium discoidale (d'Orbigny), Andersen, 1961, Dept. Conserv., La. Geol. Survey, Bull. 35, pt. 2, p. 109, pl. 18, figs. 7a, b.

Plate 35, Figure 1. Lateral view showing involute planispiral coil, densely perforate chambers, imperforate retral bands with large fossettes, and imperforate umbo containing large canal openings. USNM 304599

Plate 36, Figure 1a. Edge view showing lenticular outline, large imperforate umbos, densely perforate chambers, sparsely perforate margin, imperforate apertural face with lipped areal apertures, and cribrate basal aperture. USNM 304600

Plate 36, Figure 1b. Enlargement of apertural area showing areal and basal apertural openings and sparse perforations in peripheral margin. USNM 304600

Distribution: Formas *typicum* and *translucens* are predominant forms of genus in *Elphidium* and *Elphidium-Hanzawaia* predominance facies on Texas Shelf, where they may form a relict accumulation; elsewhere, ranges in varied abundance across continental shelf and extends into bays and estuaries; forma *typicum* is thick-walled ecophenotype resulting from delayed reproductive maturation in unfavorable environmental conditions (Poag, 1978). Illustrated specimens from Texas-Louisiana Shelf edge.

ELPHIDIUM FIMBRIATULUM (Cushman)

Polystomella fimbriatulum Cushman, 1918, U.S. Geol. Survey Bull. 676, p. 20, pl. 8, figs. 5a, b.

Elphidium cf. *E. fimbriatulum* (Cushman), Phleger and Parker, 1951, Geol. Soc. America, Mem. 46, pt. II, p. 10, pl. 5, fig. 12.
Elphidium advenum (Cushman), Parker, 1954, Mus. Comp. Zool., Harvard, Bull., vol. 111, no. 10, p. 508, pl. 6, fig. 14; Bock, 1971, Miami Geol. Soc., Mem. 1, p. 56, pl. 20, figs. 7, 8.
Elphidium fimbriatulum (Cushman), Andersen, 1961, Dept. Conserv., La. Geol. Survey, Bull. 35, pt. 2, p. 110, pl. 18, figs. 9a, b.

Plate 37, Figure 3. Lateral view showing planispiral coil, narrow retral processes with wide intervening fossettes opening into internal canal system, raised imperforate radiating ridges on adumbilical half of all but last two chambers, thick imperforate peripheral carina, pustulose and spinose surfaces around all depressions and on apertural face of final chamber. USNM 304601

Plate 38, Figure 3a. Edge view showing acutely angular periphery, thick carina, flattened apertural face of final chamber, and spinules surrounding multiple tubular areal and basal apertural openings. USNM 304602

Plate 38, Figure 3b. Enlargement of apertural region showing tubular lips surrounding basal and areal apertures, thick pustulose imperforate carina, and fine perforations in final chamber at upper right. USNM 304602

Distribution: Low to moderate numbers chiefly on inner and middle shelf; seems to be especially associated with shallow siliciclastic banks and with carbonate-rich substrates that characterize Miliolid predominance facies of Campeche and West Florida shelves, and *Amphistegina* predominance facies of submerged banks on Texas-Louisiana and West Florida shelves. Illustrated specimens from Texas Shelf.

ELPHIDIUM GALVESTONENSE Kornfeld forma *MEXICANUM*

Elphidium incertum (Williamson) var. *mexicanum* Kornfeld, 1931, Contr. Dept. Geol., Stanford Univ., vol. 1, no. 3, p. 89, pl. 16, figs. 1, 2.
cf. *Elphidium koeboeense* LeRoy, 1939, Natuurk. Tijdschr. Nedrl.-Indie, dl. 99, afl. 6, p. 240, pl. 9, figs. 6, 7.
Elphidium sp. cf. *E. koeboeense* LeRoy, Parker, Phleger, and Peirson, 1953, Cushman Found. Foram. Res., Spec. Publ. 2, p. 8, pl. 3, figs. 22, 23.
Elphidium galvestonense Kornfeld forma *mexicanum*, Poag, 1978, Gulf Coast Assoc. Geol. Socs., Trans., vol. 28, p. 403, pl. 3, figs. 1, 2, 4-11, 17, 18.

Plate 35, Figure 4. Lateral view showing planispiral coil, raised imperforate umbonal boss, wide imperforate retral processes, narrow fossettes opening into internal canal system, large pustules surrounding umbilical boss and base of final chamber, and finely perforate chamber walls. USNM 304603.

Plate 36, Figure 4a. Edge view showing rounded noncarinate periphery, raised imperforate umbilical bosses, multiple tubular basal apertures, slitlike fossettes opening into internal canal system, and finely perforate chambers. USNM 304604

Plate 36, Figure 4b. Enlargement of apertural region showing pustulose tubular basal apertures, surrounding spines, dense distribution of fine spines across periphery, and fine wall perforations. USNM 304604

Distribution: General distribution of forma is same as that of *E. galvestonense* forma *typicum*; however, this small, few-chambered, thin-walled ecophenotype is most abundant in environments providing most favorable conditions for species and thereby causing accelerated reproductive maturation (Poag, 1978). Illustrated specimens from San Antonio Bay, Texas.

ELPHIDIUM GALVESTONENSE Kornfeld forma *TYPICUM*

Elphidium gunteri Cole var. *galvestonensis* (part) Kornfeld, 1931, Contr. Dept. Geol., Stanford Univ., vol. 1, no. 3, p. 87, pl. 15, figs. 1a, b (not 2a, b; 3a, b; not of Phleger and Parker, 1951, Geol. Soc. America, Mem. 46, pt. II, p. 10, pl. 5, figs. 13, 14).
Elphidium galvestonense Kornfeld, Parker, Phleger, and Peirson, 1953, Cushman Found. Foram. Res., Spec. Publ. 2, p. 7, pl. 3, figs. 15, 16.
Elphidium galvestonense Kornfeld forma *typicum*, Poag, 1978, Gulf Coast Assoc. Geol. Socs., Trans., vol. 28, p. 403, pl. 3, figs. 3, 12-16, 19-21.

Plate 35, Figure 3. Lateral view showing scalloped outline, wide imperforate retral processes, wide fossettes opening into internal canal system, large imperforate umbilical boss, heavy pustules around umbilical boss and base of final chamber, and finely perforate chamber walls. USNM 304605

Plate 36, Figure 3a. Edge view showing thickly swollen umbilical bosses, rounded noncarinate periphery, and multiple pustulose tubular apertures at base of final chamber. USNM 304606

Plate 36, Figure 3b. Enlargement of apertural region showing partly obscured tubular apertures along chamber base, surrounding pustules and spines, and fine perforations on chamber wall at lower left. USNM 304606

Distribution: Abundant and widespread in coastal lagoons, bays, estuaries, and inner shelf environments all around Gulf; especially abundant in *Ammonia-Elphidium* predominance facies of West Texas and Mexican shelves; forma *typicum* is large, many-chambered, thick-walled ecophenotype resulting from delayed reproductive maturation in least favorable environmental conditions (Poag, 1978). Illustrated specimens from San Antonio Bay, Texas.

ELPHIDIUM GUNTERI Cole forma *SALSUM*

Cribroelphidium salsum Cushman and Brönnimann, 1948, Cushman Lab. Foram. Res., Contr., vol. 24, pt. 1, p. 19, pl. 4, fig. 6.
Elphidium delicatulum Bermudez, 1949, Cushman Lab. Foram. Res., Spec. Publ. 25, p. 168, pl. 11, figs. 22, 23; Parker, Phleger, and Peirson, 1953, Cushman Found. Foram. Res., Spec. Publ. 2, p. 7, pl. 3, figs. 12, 17.
Elphidium gunteri Cole forma *salsum*, Poag, 1978, Gulf Coast Assoc. Geol. Socs., Trans., vol. 28, p. 402, pl. 2, figs. 1-12, 17-21.

Plate 37, Figure 2. Side view showing planispiral coiling, coarsely perforate chambers, imperforate retral processes that often only partly bridge deeply depressed sutures, walls of chambers forming perforate bands wider than intervening imperforate retral bands, and umbilicus filled with spinose plugs. USNM 304607

Plate 38, Figure 2a. Side view showing flattened imperforate face of final chamber occupied by multiple apertures with spinose necks; sutural excavations do not cross periphery. USNM 304608

Plate 38, Figure 2b. Enlargement of aperture showing multiple openings surrounded by spinose necks, and perforate chamber walls below aperture. USNM 304608

Distribution: Same general distribution as *E. gunteri* forma *typicum*, but small, few-chambered, thin-walled forma *salsum* is predominant form of species in most favorable environments causing accelerated reproductive maturity (Poag, 1978). Illustrated specimens from San Antonio Bay, Texas.

ELPHIDIUM GUNTERI Cole forma *TYPICUM*

Elphidium gunteri Cole, 1931, Fla. Geol. Survey, Bull. 6, p. 34, pl. 4, figs. 9, 10; Parker, Phleger, and Peirson, 1953, Cushman Found. Foram. Res., Spec. Publ. 2, p. 8, pl. 3, figs. 18, 19.

Elphidium gunteri Cole var. *galvestonensis* Kornfeld (part), 1931, Contr. Dept. Geol., Stanford Univ., vol. 1, no. 3, p. 87, pl. 15, figs. 2a, b; 3a, b; (not figs. 1a, b).
Elphidium gunteri Cole forma *typicum*, Poag, 1978, Gulf Coast Assoc. Geol. Socs., Trans., vol. 28, p. 402, pl. 2, figs. 13-16, 22-23.

Plate 37, Figure 1. Side view showing nearly circular periphery, imperforate nodose retral processes separated by rectangular fossettes, spinose umbilical plugs, and heavily spinose wall beneath apertural face of final chamber; walls of chambers form perforate patches approximately equal in width to intervening imperforate retral bands. USNM 304609

Plate 38, Figure 1a. Edge view showing flattened face of final chamber containing multiple spinose tubular apertural openings; note nearly equal width of perforate and imperforate bands along periphery. USNM 304610

Plate 38, Figure 1b. Enlargement of aperture showing multiple spinose tubular openings and small clumps of spines on chamber beneath basal row of apertures. USNM 304610

Distribution: Abundant in coastal lagoons, bays, and estuaries around Gulf; also in lower frequencies in inner shelf facies; forma *typicum* is large, many-chambered, thick-walled ecophenotype produced by delayed reproductive maturation in least favorable environments (Poag, 1978). Illustrated specimens from San Antonio Bay, Texas.

ELPHIDIUM LANIERI (d'Orbigny)

Polystomella lanieri d'Orbigny, 1839, Foram. de Cuba, vol. 8, p. 54, pl. 7, figs. 12, 13.
Polystomella sagra d'Orbigny, 1839, Foram. de Cuba, vol. 8, p. 55, pl. 6, figs. 19, 20.
Elphidium sagrum (d'Orbigny), Bock, 1971, Miami Geol. Soc., Mem. 1, p. 56, pl. 20, figs. 11, 13.
Elphidium lanieri (d'Orbigny), Wantland, 1975, Amer. Assoc. Petrol. Geol., Stud. Geol., No. 2, p. 395, fig. 10q.

Plate 37, Figure 4. Lateral view showing planispiral coil, ovate outline, fused retral processes forming continuous ridges concentrically encircling test, ovate fossettes opening into internal canal system, fossettes wider than intervening retral processes, and densely clustered spines on early part of whorl, on flattened apertural face, and on rims of fossettes. USNM 304611

Plate 38, Figure 4a. Edge view showing inflated lenticular outline, densely spinose apertural face and ridges below final chamber, and areal apertural openings. USNM 304612

Plate 38, Figure 4b. Enlargement of apertural region showing multiple apertures in flattened face and at base of final chamber, double- and triple-pronged spines along ridges on early chambers, and single-pronged spines on apertural face. USNM 304612

Distribution: Rare in siliciclastic biotopes; low numbers in carbonate-rich areas of Campeche and West Florida shelves, and on submerged carbonate banks of northern Gulf. Illustrated specimens from central Campeche Shelf.

ELPHIDIUM POEYANUM (d'Orbigny)

Polystomella poeyana d'Orbigny, 1839, Foram. de Cuba, vol. 8, p. 55, pl. 6, figs. 25, 26.
Elphidium poeyanum (d'Orbigny), Parker, Phleger, and Peirson, 1953, Cushman Found. Foram. Res., Spec. Publ. 2, p. 9, pl. 3, fig. 26; Bock, 1971, Miami Geol. Soc., Mem. 1, p. 57, pl. 21, figs. 1, 2.

Plate 39, Figure 3. Lateral view showing planispiral coil, wide chambers, narrow imperforate retral processes, intervening subcircular fossettes opening into internal canal system, retral processes and fossettes of nearly equal width, rather uniformly distributed coarse perforations of chamber walls, and slightly depressed umbilicus covered by overlapping imperforate umbilical extensions of final few chambers. USNM 304613

Plate 40, Figure 3a. Edge view showing inflated chambers and rounded perforate periphery, subcircular fossettes, nearly imperforate patch on final chamber face, multiple basal apertures, and imperforate umbilical extensions of final chamber. USNM 304614

Plate 40, Figure 3b. Enlargement of apertural region showing subcircular fossettes, uniformly scattered coarse wall perforations, multiple basal apertures, fine spinules within apertures, and imperforate apertural face of final chamber; larger openings at extremities of aperture are typical of species. USNM 304614

Distribution: Common in coastal lagoons, bays, and estuaries, and less numerous in inner shelf facies. Illustrated specimens from San Antonio Bay, Texas.

Genus *EPISTOMINELLA* Husezima and Maruhasi, 1944

EPISTOMINELLA EXIGUA (H. B. Brady)

Pulvinulina exigua H. B. Brady, 1884, Rept. Voy. *Challenger*, Zool., vol. 9, p. 696, pl. 103, figs. 13, 14.
Epistominella exigua (H. B. Brady), Parker, 1954, Mus. Comp. Zool., Harvard, Bull., vol. 111, no. 10, p. 533, pl. 10, figs. 22, 23.

Plate 5, Figure 4. Lateral view showing subcircular shape with protruding final chamber, scattered fine perforations of chamber wall, indistinct imperforate sutures, and apertural opening parallel to margin of final chamber. USNM 304615

Plate 6, Figure 4a. Spiral view showing subcircular outline with protruding final chamber, scattered perforations, and depressed imperforate sutures. USNM 304615

Plate 6, Figure 4b. Enlargement of aperture showing serrate lip along inner right margin of aperture. USNM 304615

Plate 6, Figure 4c. Edge view showing unequally biconvex shape, low spire, scattered fine perforations, and narrow elongate aperture parallel to margin of final chamber. USNM 304615

Distribution: Persistent low numbers on middle and lower slope; scattered occurrences from outer shelf to Sigsbee Plain. Illustrated specimen from Sigsbee Plain.

EPISTOMINELLA VITREA Parker

?*Pseudoparella* cf. *exigua* (H. B. Brady, Phleger and Parker, 1951, Geol. Soc. America, Mem. 46, pt. II, p. 28, pl. 15, figs. 7a, b.
Epistominella vitrea Parker, 1953, Cushman Found. Foram. Res., Spec. Publ. 2, p. 9, pl. 4, figs. 34-36, 40, 41.

Plate 5, Figure 3. Umbilical view showing final whorl of nearly equal-sized chambers, lobate periphery, nearly straight depressed radiating sutures, sparsely perforate chamber walls, elongate aperture in depression subparallel to periphery of final chamber, aperture completely surrounded by raised serrate lip, and final chamber overlapping depressed umbilicus. USNM 304616

Plate 6, Figure 3a. Spiral side showing lobate outline, curved depressed sutures of final whorl, imperforate sutural bands of early whorl, and densely perforate chamber walls. USNM 304617

Plate 6, Figure 3b. Edge view showing biconvex outline, depressed umbilicus, imperforate depressed sutures, elongate aperture subparallel to periphery of final chamber, and serrate apertural lip. USNM 304616

Distribution: Predominant species in narrow band of *Epistominella* predominance facies adjacent to western side of Mississippi Delta; abundant elsewhere on inner and middle shelf; low numbers in bays and estuaries, and as deep as Sigsbee Plain. Illustrated specimens from Louisiana Shelf.

Genus *EPONIDES* de Montfort, 1808

EPONIDES POLIUS Phleger and Parker

Eponides polius Phleger and Parker, 1951, Geol. Soc. America, Mem. 46, pt. II, p. 21, pl. 11, figs. 1a, b; 2a, b.

Plate 1, Figure 1. Umbilical side showing lobate periphery, fine perforations, indistinct depressed radiating sutures, aperture occupying most of chamber base from umbilicus to periphery, narrow scalloped apertural flap, and faint curved ridges radiating from beneath apertural flap. USNM 304618

Plate 2, Figure 1a. Edge view showing inflated lenticular outline, imperforate sutures, elongate low aperture with arch at peripheral end, aperture occupying most of base of final chamber, narrow apertural flap, and flattened apertural face of final chamber. USNM 304619

Plate 2, Figure 1b. Spiral view showing lobate periphery, finely perforate chamber walls, and imperforate depressed sutures. USNM 304620

Distribution: Low numbers on middle and lower slope, rise, Mississippi Fan, and Sigsbee and Florida plains, especially in *Eponides-Nuttallides* predominance facies; copredominant in *Eponides* predominance facies near DeSoto Canyon. Illustrated specimens from Sigsbee Plain.

EPONIDES REGULARIS Phleger and Parker

Eponides regularis Phleger and Parker, 1951, Geol. Soc. America, Mem. 46, pt. II, p. 21, pl. 11, figs. 3a, b; 4a-c.

Plate 1, Figure 3. Umbilical view showing nearly circular periphery, numerous perforate chambers, broad imperforate sutural bands joining around deep cylindrical umbilicus, and smooth-lipped aperture approximately halfway between umbilicus and periphery. USNM 304621

Plate 2, Figure 3a. Edge view showing flattened spiral side, angular periphery, especially angular final chamber, and symmetrically arched lipped aperture located midway between umbilicus and periphery. USNM 304622

Plate 2, Figure 3b. Spiral view showing perforate chambers, nearly circular outline, and slight protrusion of early coil. USNM 304623

Distribution: Scattered occurrences in low numbers from inner shelf to Sigsbee Plain; most frequently recorded in moderate to low numbers on outer shelf and upper to middle slope. Illustrated specimens from west Campeche Slope.

EPONIDES TUMIDULUS (H. B. Brady)

Truncatulina tumidula H. B. Brady, 1884, Rept. Voy. *Challenger*, Zool., vol. 9, p. 666, pl. 95, figs. 8a-d.
Eponides tumidulus (H. B. Brady), Parker, 1954, Mus. Comp. Zool., Harvard, Bull., vol. 111, no. 10, p. 529, pl. 9, figs. 19, 24.

Plate 1, Figure 2. Umbilical view showing lobate periphery, depressed straight radiating sutures, chambers of nearly equal size, deep, narrow umbilicus, and arched serrate lip over interiomarginal aperture; perforations of wall not visible at this magnification. USNM 304624

Plate 2, Figure 2a. Edge view showing moderately high spire, inflated chambers, broadly rounded periphery, deep umbilicus, and serrate lip over arched aperture. USNM 304625

Plate 2, Figure 2b. Spiral view showing lobate periphery, broadly depressed spiral suture, and nearly equal-sized chambers of final whorl. USNM 304626

Distribution: Characteristic of and nearly restricted to *Nuttallides* and *Eponides-Nuttallides* predominance facies of lower slope, rise, and Sigsbee Plain, where it is often major constituent of assemblages; lower numbers in *Cibicides* predominance facies of Mississippi Fan. Illustrated specimens from Sigsbee Plain.

EPONIDES TURGIDUS Phleger and Parker

Eponides turgidus Phleger and Parker, 1951, Geol. Soc. America, Mem. 46, pt. II, p. 22, pl. 11, figs. 9a, b.

Plate 1, Figure 4. Umbilical view showing four quadrate chambers, quadrate test outline, straight depressed sutures that meet at nearly right angles, widely scattered pores in pits, aperture about halfway between umbilicus and periphery and covered by scalloped flap, and pustulose wall beneath aperture. USNM 304627

Plate 2, Figure 4a. Edge view showing high-spired conical shape, inflated chambers, scattered pores more frequent on umbilical side, some pores slightly rimmed, elongate aperture covered by wide flap, and pustulose wall below apertural flap. USNM 304628

Plate 2, Figure 4b. Spiral view showing subquadrate test outline, indistinct chambers of early spire, and widely scattered infrequent perforations, some with raised rims. USNM 304629

Distribution: Characteristic of *Nuttallides* and *Eponides-Nuttallides* predominance facies of rise and Sigsbee and Florida plains, where it is often major constituent of assemblages; copredominant in *Eponides* predominance facies near DeSoto Canyon; lower numbers in *Cibicides* predominance facies of Mississippi Fan; reported in low numbers from inner shelf to lower slope. Illustrated specimens from Sigsbee Plain.

Genus *FURSENKOINA* Loeblich and Tappan, 1961

FURSENKOINA COMPRESSA (Bailey)

Bulimina compressa Bailey, 1851, Smithsonian Contr., vol. 2, art. 3, p. 12, pl. 12, figs. 35-37.
Virgulina compressa (Bailey), Phleger and Parker, 1951, Geol. Soc. America, Mem. 46, pt. II, p. 19, pl. 9, figs. 4a, b; 5a, b.

Plate 33, Figure 6. Lateral view showing twisting triserial to biserial overlapping chambers, thin early third of test, fine perforations, and distinct depressed sutures. USNM 304630

Plate 34, Figure 6a. Apertural view showing finely perforate wall, broadly ovate outline, narrow aperture, and thickened apertural rim that recurves to join serrate internal tooth-plate. USNM 304631

Plate 34, Figure 6b. Edge view showing finely perforate wall, thin early test, distinct depressed sutures, narrow aperture, and thickened apertural lip that recurves to join serrate internal tooth-plate. USNM 304630

Distribution: Prominent member of *Fursenkoina* predominance facies; also abundant in other shelf facies, especially around Mississippi Delta; elsewhere in low numbers from inner shelf (including bays and estuaries) to lower slope. Illustrated specimens from East Mexico Shelf.

FURSENKOINA MEXICANA (Cushman)

Virgulina mexicana Cushman, 1922, U.S. Nat. Mus., Bull. 104, pt. 3, pl. 21, fig. 8; Parker, 1954, Mus. Comp. Zool., Harvard, Bull., vol. 111, no. 10, p. 512, pl. 7, figs. 7, 8.

Plate 33, Figure 7. Side view showing elongate overlapping chambers, sharply tapered early test, fine perforations, and elongate, narrow, nearly vertical basal aperture. USNM 304632

Plate 34, Figure 7a. Apertural view showing subcircular outline, imperforate area surrounding aperture, and narrow aperture with thin tooth-plate. USNM 304633

Plate 34, Figure 7b. Opposite side view showing alternating biserial chamber arrangement, sharply tapered early test, and faint perforations. USNM 304634

Distribution: Moderate abundances from upper slope of eastern Gulf; low abundances elsewhere from outer shelf to Sigsbee Plain. Illustrated specimens from East Mexico Shelf.

FURSENKOINA PONTONI (Cushman)

Virgulina pontoni Cushman, 1932, Cushman Lab. Foram. Res., Contr., vol. 8, pt. 1, p. 17, pl. 3, fig. 7; Parker, 1954, Mus. Comp. Zool., Harvard, Bull., vol. 111, no. 10, p. 513, pl. 7, fig. 9.

Plate 33, Figure 5. Lateral view showing twisting coil of greatly overlapping chambers, thickened early test, deeply depressed sutures, and narrow longitudinal ridges running between rows of fine perforations. USNM 304635

Plate 34, Figure 5a. Apertural view showing ovate outline, perforate wall, narrow aperture, and thickened lip at right recurving to join serrate internal tooth-plate. USNM 304636

Plate 34, Figure 5b. Edge view showing greatly overlapping chambers, thickened early test, perforate wall with narrow longitudinal ridges, distinct sutures, and narrow aperture with thickened lip on right side that recurves to join serrate internal tooth-plate. USNM 304637

Distribution: Abundant in *Fursenkoina* predominance facies of western Texas Shelf; fewer numbers in other shelf facies around Gulf. Illustrated specimens from Texas Shelf.

FURSENKOINA PUNCTATA (d'Orbigny)

Virgulina punctata d'Orbigny, 1839, Foram. de Cuba, vol. 8, p. 139, pl. 1, figs. 35, 36; Parker, 1954, Mus. Comp. Zool., Harvard, Bull., vol. 111, no. 10, p. 513, pl. 7, fig. 11.
Fursenkoina mexicana (Cushman), Bock, 1971, Miami Geol. Soc., Mem. 1, p. 62, pl. 23, fig. 8.

Plate 33, Figure 8. Side view showing nearly symmetrical lenticular outline, twisted biserial chamber arrangement, perforations of chamber walls, imperforate longitudinal ridges separating perforations in early chambers, slitlike terminal aperture crossing apex of final chamber, and strongly serrate tooth-plate. USNM 304638

Plate 34, Figure 8a. Apertural view showing strong compression of test, finer perforations and lack of ridging on final chamber, elongate aperture, thick imperforate lips, and strongly serrate tooth-plate. USNM 304639

Plate 34, Figure 8b. Edge view showing nearly parallel sides of compressed test, multiple ridging of early chambers, coarse perforations, narrow terminal aperture, and serrate tooth-plate. USNM 304640

Distribution: Distribution poorly known due to sparsity of previous records, but appears to occur most frequently in low numbers on shelf and upper slope of northeastern Gulf. Illustrated specimens from East Mexico Shelf.

Genus *GLOMOSPIRA* Rzehak, 1885

GLOMOSPIRA CHAROIDES (Jones and Parker)

Trochammina squamata var. *charoides* Jones and Parker, 1860, Quart. Jour. Geol. Soc. London, vol. 16, p. 304.
Glomospira charoides (Jones and Parker), Parker, 1954, Mus. Comp. Zool., Harvard, Bull., vol. 111, no. 10, p. 485, pl. 1, fig. 14.

Plate 7, Figure 4. Edge view showing turbinate spirally enrolled single chamber and walls composed of fine terrigenous mineral grains covered with organic film; granular structure of wall exposed where organic film has been eroded at left center and around aperture at lower center. USNM 304641

Plate 8, Figure 4a. Adapertural view showing spiral coiling, aperture, and granular structure of walls where organic surface film has been eroded on outermost whorl of chamber. USNM 304641

Plate 8, Figure 4b. Abapertural view showing spiral coiling, organic film covering walls, and subcircular outline. USNM 304641

Distribution: Characteristic and sometimes predominant on elevated prominences of slope, rise, Sigsbee Plain, and Mississippi Fan, lending its name to agglutinant *Glomospira* predominance facies; scattered occurrences in low numbers also reported from middle and outer shelf. Illustrated specimen from East Mexico Slope.

Genus "*GOESELLA*"

"*GOESELLA*" *MISSISSIPPIENSIS* Parker

Goesella mississippiensis Parker, 1954, Mus. Comp. Zool., Harvard, Bull., vol. 111, no. 10, p. 495, pl. 3, figs. 13, 14, 19.

Plate 39, Figure 4. Side view showing triserial coiling, lunate terminal aperture, and wall composed of angular terrigenous mineral grains arranged with flat sides outward in finer grained matrix. USNM 304642

Plate 40, Figure 4a. Apertural view showing lunate, nontoothed aperture, subcircular test outline, and coarse grains of final chamber. USNM 304643

Plate 40, Figure 4b. Prolocular view showing triserial chamber arrangement, subcircular outline, and coarse-grained composition of wall. USNM 304644

Distribution: Known only from vicinity of Mississippi Delta, where it is abundant and predominant in "*Goesella*" predominance facies of middle shelf to upper slope; fewer numbers on shelf and slope off Mississippi and Alabama; morphology, especially apertural structure, does not conform to genus *Goesella* Cushman, as described by Loeblich and Tappan (1964); appears to warrant erection of new genus. Illustrated specimens from Louisiana Slope.

Genus *HANZAWAIA* Asano, 1944

HANZAWAIA CONCENTRICA (Cushman) forma *STRATTONI*

Truncatulina americana Cushman var. *strattoni* Applin, 1925, Amer. Assoc. Petrol. Geol., Bull., vol. 9, no. 1, p. 99, pl. 3, fig. 3.
Cibicidina strattoni (Applin), Parker, Phleger, and Peirson, 1953, Cushman Found. Foram. Res., Spec. Publ. 2, p. 7, pl. 4, figs. 38, 39.

Plate 39, Figure 2. Umbilical view showing uniformly perforate chamber walls, thick raised imperforate sutures of early chambers, coalescing imperforate apertural flaps forming concentric band around umbilicus, and previous apertures remaining partly open at outer edge of umbilical band. USNM 304645

Plate 40, Figure 2a. Edge view showing rounded periphery of inflated final chamber, slightly acuminate periphery of earlier chambers, imperforate face of final chamber, imperforate sutures and peripheral band in early chambers, and lipped aperture crossing periphery. USNM 304646

Plate 40, Figure 2b. Umbonal view showing uniform perforations, thick raised sutures between all but last four chambers, depressed umbo and arched apertural lip crossing periphery. USNM 304647

Distribution: Distribution identical with that of forma *typica*; forma *strattoni* somewhat larger, more inflated, and less angular peripherally than forma *typica* and appears to be ecophenotypic variant. Illustrated specimens from Texas Shelf.

HANZAWAIA CONCENTRICA (Cushman) forma *TYPICA*

Truncatulina concentrica Cushman, 1918, U.S. Geol. Survey, Bull. 676, p. 64, pl. 21, fig. 3.
Cibicides concentricus (Cushman), Phleger and Parker, 1951, Geol. Soc. America, Mem. 46, pt. II, p. 29, pl. 15, figs. 14a, b; (?15a, b).

Plate 39, Figure 1. Umbilical view showing uniform perforations of chamber walls, thick raised imperforate sutures and carina, imperforate apertural flaps coalescing to form concentric band around umbilicus, and former apertures remaining partly open around outer edge of umbilical band. USNM 304648

Plate 40, Figure 1a. Edge view showing acute periphery and imperforate carina, arched apertural lip crossing periphery, and depressed umbo. USNM 304649

Plate 40, Figure 1b. Umbonal view showing uniform perforations of chamber walls, thick raised imperforate sutures, peripheral carina, and umbonal boss. USNM 304650

Distribution: Common in moderate to large numbers on inner and middle shelf throughout Gulf; also occurring in bays and estuaries; formae *typica* and *strattoni* are predominant or co-predominant in *Hanzawaia*, *Hanzawaia-Planulina*, *Elphidium-Hanzawaia*, and *Rosalina-Hanzawaia* predominance facies; occasional specimens reported from slope are probably displaced; forma *typica* is thicker walled and more sharply keeled than forma *strattoni*; these morphologic differences appear to be environmentally controlled; no other species of *Hanzawaia* known in Gulf. Illustrated specimens from Texas Shelf.

Genus *HAPLOPHRAGMOIDES* Cushman, 1910

HAPLOPHRAGMOIDES BRADYI (Robertson)

?*Trochammina robertsoni* H. B. Brady, 1887 (not *T. robertsoni* H. B. Brady, 1876), Jour. Roy. Micr. Soc., p. 893.
?*Trochammina bradyi* Robertson, 1891, Ann. Mag. Nat. Hist., Ser. 6, vol. 7, p. 388.
Haplophragmoides bradyi (Robertson), Parker, 1954, Mus. Comp. Zool., Harvard, Bull., vol. 111, no. 10, p. 486, pl. 1, fig. 16.

Plate 13, Figure 4. Side view showing planispiral coiling, straight radiating sutures, and test wall composed of large flattened terrigenous mineral grains set in finer grained matrix and covered with thin, smooth organic film; organic film eroded in several spots on final chambers, exposing underlying particles. USNM 304651

Plate 14, Figure 4a. Enlargement of aperture showing well-sorted fine particles of aperture and poorly sorted larger particles of wall beneath aperture; organic film obscures angular edges of mineral grains. USNM 304651

Plate 14, Figure 4b. Edge view showing ovate shape, slightly depressed umbilici, symmetrical basal aperture, particulate wall, and organic film missing over large portion of final chamber, exposing sharp edges of mineral grains. USNM 304651

Distribution: Common to abundant in some parts of agglutinant *Glomospira* predominance facies of slope, rise, and Sigsbee Plain; reported in fewer numbers from inner to outer shelf. Illustrated specimen from Sigsbee Plain.

Genus *HOEGLUNDINA* Brotzen, 1948

HOEGLUNDINA ELEGANS (d'Orbigny)

Rotalia (Turbulina) elegans d'Orbigny, 1826, Ann. Sci. Nat., vol. 7, p. 276, no. 54.
Hoeglundina elegans (d'Orbigny), Parker, 1954, Mus. Comp. Zool., Harvard, Bull., vol. 111, no. 10, p. 531, pl. 10, figs. 4, 8.

Plate 19, Figure 3. Umbilical view showing nearly circular outline, smooth imperforate test surface, slitlike aperture near periphery of final chamber, and depressions along periphery representing previous apertures. USNM 304652

Plate 20, Figure 3a. Spiral view showing nearly circular outline and smooth imperforate test surface. USNM 304653

Plate 20, Figure 3b. Edge view showing lenticular outline, smooth imperforate test surface, keeled periphery, and elongate aperture on umbilical side of periphery. USNM 304654

Plate 20, Figure 3c. Apertural enlargement showing imperforate test surface and scattered short spinules on abumbilical wall of aperture. USNM 304654

Distribution: Scattered occurrences in low numbers from inner shelf to Sigsbee Plain; isolated concentrations along lower slope and on rise in northeastern Gulf form *Hoeglundina* predominance facies; only species of *Hoeglundina* known in Gulf. Illustrated specimens from Louisiana Slope.

Genus *HOMOTREMA* Hickson, 1911

HOMOTREMA RUBRA (Lamarck)

Millipora rubra Lamarck, 1816, Hist. Nat. Anim. sans Vert., vol. 2, p. 202.
Polytrema rubra Dujardin, 1841, Hist. Nat. Zooph., p. 259.
Homotrema rubrum Hickson, 1911, Trans. Linn. Soc. London, Zool., Ser. 2, vol. 14, p. 445, 454; pl. 30, fig. 2; pl. 31, fig. 9; pl. 32, figs. 19, 22, 28; Bock, 1971, Miami Geol. Soc., Mem. 1, p. 61, pl. 23, fig. 3.

Plate 45, Figure 4. Lateral view showing conical branching form, dense coarse perforations, and multiple branching tubular chambers. USNM 304655

Plate 46, Figure 4a. Cross-section view showing coarse, dense perforations of outer wall, and subconcentric layers of tubular chambers. USNM 304656

Plate 46, Figure 4b. Lateral view of branch tip showing tubular form, dense, coarse perforations, and imperforate sutural ridges. USNM 304657

Distribution: Common as fragments in carbonate-rich provinces such as West Florida and Campeche shelves and submerged banks of northern Gulf; most abundant on living emergent reefs of Campeche Shelf, where it is one of three predominant forms in Miliolid-*Archaias-Homotrema* predominance facies; only species of *Homotrema* known in Gulf. Illustrated specimens from West Flower Garden Bank, Texas Shelf.

Genus *ISLANDIELLA* Nørvang, 1958

ISLANDIELLA CURVATA (Phleger and Parker)

Cassidulina curvata Phleger and Parker, 1951, Geol. Soc. America, Mem. 46, pt. II, p. 26, pl 14, figs. 5a, b.

Plate 17, Figure 1. Lateral view showing crescentic coarsely perforate chambers, curved imperforate sutural depressions, imperforate and somewhat serrate carina, aperture near periphery, with lip and tooth-plate raised above apertural depression. USNM 304658

Plate 18, Figure 1a. Edge view showing lenticular outline of test, sweeping curve of aligned perforations paralleling sutures, narrow serrate imperforate carina, and narrow peripheral aperture with serrate tooth-plate. USNM 304659

Plate 18, Figure 1b. Enlargement of aperture showing narrow, internal lip at right, which folds back on itself at each end and joins serrate tooth-plate. USNM 304659

Distribution: Common in moderate to large numbers on shelf and upper slope around Gulf; fewer numbers from Sigsbee Plain. Illustrated specimens from Texas-Louisiana Shelf edge.

ISLANDIELLA LAEVIGATA (d'Orgigny)

Cassidulina laevigata d'Orbigny, 1826, Ann. Sci. Nat., vol. 7, p. 282, no. 1, pl. 15, figs. 4, 5; Modèles, no. 41; Parker, 1954, Mus. Comp. Zool., Harvard, Bull., vol. 111, no. 10, p. 536, pl. 11, fig. 2.

Plate 17, Figure 2. Lateral view showing lobate outline, indistinct sutures, swollen umbo, and faintly reticulate periphery truncated adjacent to elongate aperture. USNM 304660

Plate 18, Figure 2a. Edge view showing swollen umbos giving lenticular shape, roundly angular periphery, peripheral aperture near center of final chamber, and internal lip recurved at both ends and joining tooth-plate. USNM 304661

Plate 18, Figure 2b. Enlargement of aperture showing internal lip at right, which recurves at both ends and joins serrate tooth-plate. USNM 304660

Distribution: Low numbers chiefly on inner, middle, and outer shelf; scattered occurrences as deep as Sigsbee Plain. Illustrated specimens from east Campeche Shelf.

ISLANDIELLA NORCROSSI AUSTRALIS (Phleger and Parker)

Cassidulina norcrossi australis Phleger and Parker, 1951, Geol. Soc. America, Mem. 46, pt. II, p. 27, pl. 14, figs. 8a, b; 9, 10.

Plate 19, Figure 2. Lateral view showing subcircular nearly smooth outline, scattered pores concentrated mainly on final chambers, indistinct sutures, and swollen umbo. USNM 304662

Plate 20, Figure 2a. Edge view showing rounded imperforate carina, lenticular test outline, scattered pores, peripheral aperture, serrate inner lip and tooth-plate. USNM 304663

Plate 20, Figure 2b. Enlargement of aperture showing serrate inner lip at right, recurving at both ends to join serrate tooth-plate. USNM 304663

Distribution: Sparse to moderate numbers on shelf and slope around Gulf. Illustrated specimens from Texas-Louisiana Shelf edge.

ISLANDIELLA SUBGLOBOSA (H. B. Brady)

Cassidulina subglobosa H. B. Brady, 1881, Quart. Jour. Micr. Sci., new ser., vol. 21, p. 60; 1884, Rept. Voy. *Challenger*, Zool., vol. 9, p. 430, pl. 54, figs. 17 a-c.
Cassidulina subglobosa H. B. Brady, Phleger and Parker, 1951, Geol. Soc. America, Mem. 46, pt. II, p. 27, pl. 14, figs. 11, 12 (not 13).
Cassidulina subglobosa H. B. Brady, Parker, 1954, Mus. Comp. Zool., Harvard, Bull., vol. 111, no. 10, p. 536, pl. 11, fig. 9 (not 4-8).
Globocassidulina mollucensis (Gemeraad), Pflum and Frerichs, 1976, p. 116, pl. 4, fig. 5.

Plate 17, Figure 3. Side view showing subspherical shape, inflated chambers, coarsely perforate chamber walls except around aperture, shallow grooves radiating from aperture, elongate aperture, crenulate outer lip, and serrate internal tooth-plate. USNM 304664

Plate 18, Figure 3a. Abapertural view showing subspherical shape and faint imperforate sutures separating coarsely perforate chambers. USNM 304665

Plate 18, Figure 3b. Apertural view showing imperforate sutures, inflated chambers, outer crenulate apertural lip, and serrate internal tooth-plate; radiating grooves around aperture somewhat obscured in this specimen. USNM 304666

Plate 18, Figure 3c. Side view showing outer crenulate lip of aperture, and grooves radiating from aperture. USNM 304667

Distribution: Abundant in *Islandiella* and *Islandiella-Rosalina* predominance facies of outer part of northern West Florida Shelf and lower slope; also common in other slope localities; rare on shelf, Mississippi Fan, and Sigsbee Plain. Illustrated specimens from middle Campeche Slope.

ISLANDIELLA sp. aff. *I. SUBGLOBOSA* (H. B. Brady)

aff. *Cassidulina subglobosa* H. B. Brady, 1881, Quart. Jour. Micr. Sci., new ser., vol. 21, p. 60; 1884, Rept. Voy. *Challenger*, Zool., vol. 9, p. 430, pl. 54, figs. 17 a-c.
Cassidulina subglobosa H. B. Brady, Phleger and Parker, 1951, Geol. Soc. America, Mem. 46, pt. II, p. 27, pl. 14, fig. 13 (not figs. 11, 12).
Cassidulina subglobosa H. B. Brady, Parker, 1954, Mus. Comp. Zool., Harvard, Bull., vol. 111, no. 10, p. 536, pl. 11, figs. 4, 5 (?6, 7, 8; not 9).
Cassidulina subglobosa H. B. Brady, Andersen, 1961, Dept. Conserv., La. Geol. Survey, Bull. 35, p. 113, pl. 25, figs. 2 a-c.

Plate 17, Figure 4. Side view showing ovate shape, enrolled biserial coiling, perforate chamber walls, deeply excavated narrow sutures, and elongate V-shaped aperture with serrate tooth-plate. USNM 304668

Plate 18, Figure 4a. Apertural view showing imperforate area around aperture and serrate tooth-plate. USNM 304669

Plate 18, Figure 4b. Enlargement of aperture showing V-shaped opening and serrate tooth-plate; upper rim of apertural lip has bifurcated to form irregular supplementary opening. USNM 304669

Plate 18, Figure 4c. Edge view showing ovate outline of test, deeply excavated sutures, V-shaped aperture, and serrate tooth-plate. USNM 304669

Distribution: Probably recorded as *Cassidulina subglobosa* by most investigators who have not used scanning electron photomicrographs to distinguish apertural characteristics; therefore, total distribution not known; low numbers on Campeche Shelf; presumably distribution is similar to that of *Islandiella subglobosa*; tooth-plate distinguishes it as *Islandiella* but is quite distinct from tooth-plate of *I. subglobosa*. Illustrated specimens from east Campeche Shelf.

ISLANDIELLA sp.

Plate 19, Figure 1. Lateral view showing biserially enrolled chambers, dense scattering of large perforations, narrow, distinct imperforate sutures, lobate subcircular outline, peripheral aperture near middle of final chamber, and serrate internal lip and tooth-plate. USNM 304670

Plate 20, Figure 1a. Edge view showing sharply angulate periphery, lenticular shape, coarse densely scattered perforations, and aperture with tooth-plate. USNM 304671

Plate 20, Figure 1b. Enlargement of aperture showing internal lip at right recurving at each end to join serrate tooth-plate. USNM 304671

Distribution: Moderately abundant on Campeche Shelf; distribution not known elsewhere because this is first record from Gulf. Illustrated specimen from east Campeche Shelf.

Genus *KARRERIELLA* Cushman, 1933

KARRERIELLA APICULARIS (Cushman)

Gaudryina apicularis Cushman, 1911, U.S. Nat. Mus., Bull., vol. 71, pt. 2, p. 69, figs. 110a, b.
Plectina apicularis (Cushman), Parker, 1954, Mus. Comp. Zool., Harvard, Bull., vol. 111, no. 10, p. 495, pl. 3, fig. 18.

Plate 15, Figure 5. Side view showing early multiserial stage and later biserial chamber arrangement, depressed sutures, apical apertural neck, and poorly sorted larger terrigenous mineral grains set in finer matrix of chamber walls. USNM 304672

Plate 16, Figure 5a. Apertural view showing circular outline, coarser grains set in finer matrix, and small subcircular necked aperture. USNM 304672

Plate 16, Figure 5b. Side view showing acuminate prolocular end, depressed sutures, and coarse, poorly sorted surface grains. USNM 304673

Distribution: Common in agglutinant *Glomospira* predominance facies on lower slope, rise, and Sigsbee Plain; rare as shallow as upper slope. Illustrated specimens from East Mexico Slope and Sigsbee Plain.

Genus *MILIOLINELLA* Wiesner, 1931

MILIOLINELLA CIRCULARIS (Bornemann)

Triloculina circularis Bornemann, 1855, Geol. Ges. Zeitschr., vol. 7, pt. 2, p. 349, pl. 19, figs. 4a-c.
Miliolinella circularis (Bornemann), Wantland, 1975, Amer. Assoc. Petrol. Geol., Stud. Geol., No. 2, p. 390, fig. 15f, g; [not *M. circularis* (Bornemann), Bock, 1971, Miami Geol. Soc., Mem. 1, p. 29, pl. 12, fig. 5 = *M. labiosa* (d'Orbigny)].

Plate 59, Figure 3. Side view showing three smooth chambers of final whorl, ovate outline, and wide flap over aperture. USNM 304674

Plate 60, Figure 3a. View opposite that of Plate 59, Figure 3. USNM 304675

Plate 60, Figure 3b. Apertural view showing ovate outline and wide flap covering aperture; upper rim of aperture abnormally split rather than smoothly arched as usual. USNM 304676

Distribution: Common to abundant in Miliolid predominance facies and other facies with abundant miliolids. Illustrated specimens from east Campeche Shelf.

MILIOLINELLA FICHTELIANA (d'Orbigny)

Triloculina fichteliana d'Orbigny, 1839, Foram. de Cuba, vol. 8, p. 171, pl. 9, figs. 8-10.
Miliolinella fichteliana (d'Orbigny), Bock, 1971, Miami Geol. Soc., Mem. 1, p. 29, pl. 12, fig. 6.

Plate 59, Figure 4. Side view showing three chambers of final whorl, broadly ribbed surface, large lipped aperture, and small apertural flap. USNM 304677.

Plate 60, Figure 4a. View opposite that of Plate 59, Figure 4, showing deeply depressed suture, broadly ribbed surface, and lipped apical aperture. USNM 304678

Plate 60, Figure 4b. Apertural view showing three chambers of final whorl, broad ribbing, and wide lipped aperture with small flap. USNM 304679

Distribution: Rare to common in Miliolid predominance facies and other facies with abundant miliolids. Illustrated specimens from east Campeche Shelf.

Genus *NODOBACULARIELLA* Cushman and Hanzawa, 1937

NODOBACULARIELLA CASSIS (d'Orbigny)

Vertebralina cassis d'Orbigny, 1839, Foram. de Cuba, vol. 8, p. 51, pl. 7, figs. 14, 15.
Vertebralina mucronata d'Orbigny, 1839, Foram. de Cuba, vol. 8, p. 52, pl. 7, figs. 16-19.
Nodobaculariella cassis (d'Orbigny), Phleger and Parker, 1951, Geol. Soc. America, Mem. 46, pt. II, p. 8, pl. 4, figs. 12-14.
Nodobaculariella atlantica Cushman and Hanzawa, Andersen, 1961, Dept. Conserv., La. Geol. Survey, Bull. 35, pt. 2, p. 44, pl. 10, fig. 5.

Plate 61, Figure 4. Lateral view showing early coil, single chamber of rectilinear portion, strongly costate chambers, distinct sutures, and wide thin peripheral carinae. USNM 304680

Plate 62, Figure 4a. Lateral view showing two chambers in rectilinear portion of test. USNM 304681

Plate 62, Figure 4b. Edge view showing narrow flattened test and flaring lips of elongate aperture. USNM 304682

Plate 62, Figure 4c. Apertural view showing everted imperforate lips and termination of peripheral carina at each end of aperture. USNM 304683

Distribution: Common to abundant in Miliolid, Miliolid-Soritid, and *Asterigerina*-Miliolid predominance facies; also common in relict *Amphistegina* facies of submerged carbonate banks in northern Gulf. Illustrated specimens from central Campeche Shelf and West Flower Garden Bank, Texas Shelf.

Genus *NONIONELLA* Cushman, 1926

NONIONELLA BASILOBA Cushman and McCulloch

Nonionella basiloba Cushman and McCulloch, 1940, Allan Hancock Pacific Exped., vol. 6, no. 3, p. 162, pl. 18, fig. 3a-c.
Nonionella opima Cushman, 1947, Cushman Lab. Foram. Res., vol. 23, pt. 4, p. 90, pl. 20, figs. 1-3.
Nonionella cf. *N. opima* Cushman, Phleger and Parker, 1951, Geol. Soc. America, Mem. 46, pt. II, p. 11, pl. 6, figs. 1a, b; 2a, b.

Plate 43, Figure 4. Side view showing ovate outline, spiral coil, depressed sutures, finely perforate walls, and large inflated final chamber overlapping umbilicus. USNM 304684

Plate 44, Figure 4a. Opposite side showing ovate outline, finely perforate walls, and depressed sutures between final two chambers. USNM 304685

Plate 44, Figure 4b. Edge view showing roundly acute periphery, faint imperforate sutures separating finely perforate chambers, and multiple tubular apertures beneath edge of overlapping final chamber. USNM 304686

Distribution: Predominant along with less inflated forms of species in *Nonionella* predominance facies of Mississippi Delta region; common to abundant elsewhere on shelf around Gulf; rare in deeper water facies. Illustrated specimens from Texas-Louisiana Shelf.

Genus *NOURIA* Heron-Allen and Earland, 1914

NOURIA POLYMORPHINOIDES Heron-Allen and Earland

Nouria polymorphinoides Heron-Allen and Earland, 1914, Trans., Zool. Soc. London, vol. 20, pt. 12,

p. 376, pl. 37, figs. 1-15; Parker, 1954, Mus. Comp. Zool., Harvard, Bull., vol. 111, no. 10, p. 503, pl. 5, fig. 19.
Proteonina comprima Phleger and Parker, 1951, Geol. Soc. America, Mem. 46, pt. II, p. 2, pl. 1, figs. 1-3.

Plate 49, Figure 2. Side view showing single chamber of large poorly sorted terrigenous mineral grains set in finer grained matrix; larger grains decrease in size toward apical aperture. USNM 304687

Plate 50, Figure 2a. View opposite that of Plate 49, Figure 2, showing slightly depressed sublongitudinal suture and poorly sorted mineral grains in wall. USNM 304688

Plate 50, Figure 2b. Apertural view showing small grains surrounding aperture and larger grains in wall below. USNM 304689

Distribution: Predominant in *Nouria* predominance facies on outer shelf around Mississippi Delta; also common in *Saccammina-Ammobaculites* predominance facies of Texas-Louisiana Shelf; elsewhere in low numbers on inner and middle shelf. Illustrated specimens from Texas Shelf.

Genus *NUTTALLIDES* Finlay, 1939

NUTTALLIDES DECORATA (Phleger and Parker)

Pseudoparella(?) *decorata* Phleger and Parker, 1951, Geol. Soc. America, Mem. 46, pt. II, p. 28, pl. 15, figs. 4a, b; 5a, b.
Epistominella decorata (Phleger and Parker), Parker, 1954, Mus. Comp. Zool., Harvard, Bull., vol. 111, no. 10, p. 533, pl. 10, figs. 18, 19.
Alabamina decorata (Phleger and Parker), Pflum and Frerichs, 1976, Cushman Found. Foram. Res., Spec. Publ. 14, p. 120, pl. 6, figs. 8, 9; pl. 7, fig. 1.

Plate 5, Figure 2. Umbilical view showing lobate periphery, slightly rugose wall, sinuate depressed sutures, large, randomly scattered rimmed pores, large, lipped, spatulate flap covering aperture, apertural wall of final chamber recurved along periphery toward earlier chambers, and spinules below apertural flap. USNM 304690

Plate 6, Figure 2a. Spiral view showing heavily crustose center of test, tangentially radiating sutures, and scattered protruding pore rims. USNM 304691

Plate 6, Figure 2b. Edge view showing biconvex outline, subangulate periphery and alabamine extension of final chamber along periphery below aperture; aperture below flap filled with sediment. USNM 304692

Distribution: Widespread in abyssal and bathyal depths but especially abundant in *Nuttallides* predominance facies, where it is predominant, and in *Eponides-Nuttallides* predominance facies of Sigsbee Plain and Western Gulf Rise, where it is copredominant; internal and external apertural structure place this species in *Nuttallides*; only species of *Nuttallides* known in Gulf; possibly indigenous. Illustrated specimens from Sigsbee Plain.

Genus *PENEROPLIS* de Montfort, 1808

PENEROPLIS PROTEUS d'Orbigny

Peneroplis protea d'Orbigny, 1839, Foram. de Cuba, vol. 8, p. 60, pl. 7, figs. 7-11.
Peneroplis carinatus d'Orbigny, 1839, Voy. Amér. Mérid., vol. 5, pt. 5, "Foraminifères", p. 33, pl. 3, figs. 7, 8.
Peneroplis bradyi Cushman, 1930, U.S. Nat. Mus., Bull. 104, pt. 7, p. 40, pl. 14, figs. 8-10.
Peneroplis proteus d'Orbigny, Phleger and Parker, 1951, Geol. Soc. America, Mem. 46, pt. II, p. 11, pl. 6, fig. 6.

Plate 47, Figure 3. Side view showing early planispiral coil and later concentric annual chambers of adult specimen; note lack of chamber partitions in broken chamber at lower right; perforations of chamber walls indistinct at this magnification. USNM 304693

Plate 48, Figure 3a. Edge view of adult specimen showing truncate edge, inflated spire of juvenile portion protruding from flattened adult portion, and multiple rimmed apertures along entire periphery. USNM 304694

Plate 48, Figure 3b. Side view of juvenile specimen showing planispiral coil, finely perforate upper parts of chambers, depressed narrow sutures, and multiple rimmed apertures on flattened face of penultimate chamber; ultimate chamber broken off. USNM 304695

Plate 48, Figure 3c. Edge view of juvenile specimen showing lenticular outline, variable density of perforations on each chamber, depressed sutures, flattened perforate face of final chamber, and multiple rimmed apertures. USNM 304696

Distribution: Common to abundant in *Archaias, Asterigerina*-Soritid, Miliolid-Soritid, Miliolid-*Archaias-Homotrema*, and Miliolid predominance facies of carbonate-rich shelf biotopes; rare in relict *Amphistegina* facies on submerged banks of northern Gulf; several species of *Peneroplis* have been reported in Gulf but, in my opinion, all belong to single species *P. proteus*. Illustrated specimens from central Campeche Shelf.

Genus *PLANULINA* d'Orbigny, 1826

PLANULINA ARIMINENSIS d'Orbigny

Planulina ariminensis d'Orbigny, 1826, Ann. Sci. Nat., vol. 7, p. 280, no. 1, pl. 14, figs. 1-3 bis.; Modèles, no. 49.
Planulina ariminensis d'Orbigny, Phleger and Parker, 1951, Geol. Soc. America, Mem. 46, pt. II, p. 32, pl. 18, figs. 4a, b.

Plate 43, Figure 3. Side view showing deeply pitted rugose perforate chamber walls, evolute coiling, deeply depressed sutures, depressed umbilicus, and narrow apertural flaps. USNM 304697

Plate 44, Figure 3a. View opposite that of Plate 43, Figure 3, showing deeply pitted, rugose chamber walls, slightly evolute coiling, and depressed umbilicus and sutures. USNM 304698

Plate 44, Figure 3b. Edge view showing angulate keeled periphery, rugose chamber walls on both sides of test, flattened imperforate apertural face of final chamber, and rimmed aperture crossing periphery. USNM 304699

Distribution: Widespread in low numbers from inner shelf to upper slope. Illustrated specimens from east Campeche Shelf.

PLANULINA EXORNA Phleger and Parker

Planulina exorna Phleger and Parker, 1951, Geol. Soc. America, Mem. 46, pt. II, p. 32, pl. 18, figs. 5a, b; 6a, b; 7a, b; 8a, b.

Plate 43, Figure 2. Side view showing deeply pitted perforate walls, peripheral carina, evolute coiling, and apertural flaps. USNM 304700

Plate 44, Figure 2a. View opposite that of Plate 43, Figure 2, showing evolute coil, raised imperforate sutures in early coil becoming depressed between final two chambers, pitted perforate chamber wall in final five chambers, and perforate walls in earlier chambers. USNM 304701

Plate 44, Figure 2b. Edge view showing imperforate band along acute subtruncate periphery, flattened apertural face of final chamber, and lipped aperture arching across periphery. USNM 304701

Distribution: One of most abundant species of genus in *Planulina* and *Hanzawaia-Planulina* predominance facies of the West Florida Shelf, where it may form a relict accumulation; also common elsewhere on inner and middle shelf. Illustrated specimens from Texas Shelf.

PLANULINA FOVEOLATA (H. B. Brady)

Anomalina foveolata H. B. Brady, 1884, Rept. Voy. *Challenger*, Zool., vol. 9, p. 674, pl. 94, figs. 1a-c.
Planulina foveolata (H. B. Brady), Phleger and Parker, 1951, Geol. Soc. America, Mem, 46, pt. II, p. 33, pl. 18, figs. 9a, b; 10a, b.

Plate 43, Figure 1. Side view showing strongly raised rugose sutures in early imperforate evolute coil and less accentuated sutures between later perforate chambers. USNM 304702

Plate 44, Figure 1a. View opposite that of Plate 43, Figure 1, showing weakly raised sutures between all but final two chambers, evolute coiling, apertural flap, and coarsely perforate chamber walls. USNM 304703

Plate 44, Figure 1b. Edge view showing truncate periphery and arched, lipped aperture crossing left rim of periphery; note contrast in morphology of two sides. USNM 304704

Distribution: Common in *Planulina* and *Hanzawaia-Planulina* predominance facies of West Florida Shelf; also moderately abundant in other shelf localities throughout Gulf; a few occurrences on upper slope also reported. Illustrated specimens from Texas-Louisiana Shelf edge.

Genus *QUINQUELOCULINA* d'Orbigny, 1826

QUINQUELOCULINA AGGLUTINANS d'Orbigny

Quinqueloculina agglutinans d'Orbigny, 1839, Foram. de Cuba, vol. 8, p. 195, pl. 2, figs. 11-13; Le Calvez, 1977, Cah. Micropaleont., no. 1, p. 54, pl. 7, figs. 1-4.
Quinqueloculina bidentata d'Orbigny, 1971, Miami Geol. Soc., Mem. 1, p. 17, pl. 4, fig. 12; pl. 5, figs. 1-2.

Plate 59, Figure 1. Lateral view showing elliptical outline, three of five chambers of final whorl, lipped terminal aperture, test surface composed of poorly sorted rounded terrigenous and biogenic mineral grains set in finer grained matrix. USNM 304705

Plate 60, Figure 1a. View opposite that of Plate 59, Figure 1. USNM 304706

Plate 60, Figure 1b. Apertural view showing subpentagonal outline, five angulate agglutinated chambers, scalloped margin of apertural lip, and elongate slightly bifid apertural tooth. USNM 304707

Distribution: Common to abundant in Miliolid predominance facies; also common in other facies with abundant miliolids; rare to absent on most siliciclastic substrates. Illustrated specimens from central Campeche Shelf.

QUINQUELOCULINA BICARINATA d'Orbigny

Quinqueloculina bicarinata d'Orbigny, 1826, Ann. Sci. Nat., Ser. 1, vol. 7, p. 302; Bock, 1971, Miami Geol. Soc., Mem. 1, p. 16, pl. 6, figs. 7-9 (not pl. 4, figs. 6-8).

Plate 55, Figure 3. Lateral view showing elliptical outline, four of five alternating chambers of final whorl, lipped terminal aperture, and faint triple carinae on early two chambers. USNM 304708

Plate 56, Figure 3a. View opposite that of Plate 55, Figure 3. USNM 304709

Plate 56, Figure 3b. Apertural view showing subtriangular outline, five angular chambers of final whorl, faint triple carinae at chamber angles, subcircular aperture with smooth lip on three sides, and narrow bifid apertural tooth. USNM 304710

Distribution: Chiefly in low to moderate numbers in Miliolid predominance facies; also in other miliolid-rich facies of West Florida Shelf. Illustrated specimens from central Campeche Shelf.

QUINQUELOCULINA BICOSTATA d'Orbigny forma *GARRETTII*

Quinqueloculina garrettii Andersen, 1961, Dept. Conserv., La., Geol. Survey, Bull. 35, pt. 2, p. 30, pl. 4, figs. 5a-c.

Plate 63, Figure 2. Lateral view showing broadly ovate outline, four of final five chambers, moderately striate surface, strong longitudinal carinae, deeply concave chamber walls, and terminal aperture with everted lip and protruding tooth. USNM 304711

Plate 64, Figure 2a. View opposite that of Plate 63, Figure 2. USNM 304712

Plate 64, Figure 2b. Apertural view showing stellate outline, strongly tricarinate final two chambers, partly truncate carinae, deeply concave chamber walls, and circular lipped aperture with bifid tooth. USNM 304713

Distribution: Rare to common in Miliolid predominance facies; absent or rare elsewhere. Illustrated specimens from central Campeche Shelf.

QUINQUELOCULINA BICOSTATA d'Orbigny forma *TYPICA*

Quinqueloculina bicostata d'Orbigny, 1839, Foram. de Cuba, vol. 8, p. 195, pl. 12, figs. 8-10; Andersen, 1961, Dept. Conserv., La. Geol. Survey, Bull. 35, pt. 2, p. 29, pl. 4, figs. 4a-c.

Plate 63, Figure 1. Lateral view showing elliptical outline, four of five final chambers, fine striations of chamber surfaces, sharp double carinae along margin of central chamber, and lipped terminal aperture. USNM 304714

Plate 64, Figure 1a. View opposite that of Plate 63, Figure 1. USNM 304714

Plate 64, Figure 1b. Apertural view showing subtriangular outline, subcircular aperture with short knobbed tooth, striations of chamber surfaces, double peripheral carinae, and concave inter-carina surfaces. USNM 304714

Distribution: Common in Miliolid predominance facies and other West Florida Shelf localities with abundant miliolids; one of most widespread species of *Quinqueloculina*, having been recorded also on inner part of Texas-Louisiana Shelf and on submerged carbonate banks of northern Gulf. Illustrated specimen from Texas Shelf.

QUINQUELOCULINA CANDEIANA d'Orbigny

Quinqueloculina candeiana d'Orbigny, 1839, Foram. de Cuba, vol. 8, p. 199, pl. 12, figs. 24-26; Le Calvez, 1977, Cah. Micropaleont., no. 1, p. 68, pl. 11, figs. 1-3.
Quinqueloculina lamarckiana d'Orbigny, Phleger and Parker, 1951, Geol. Soc. America, Mem. 46, pt. II, p. 7, pl. 4, figs. 1a, b; 2a, b.

Plate 55, Figure 4. Lateral view showing broadly ovate outline, four of final five chambers, smooth surface, deeply excavated narrow sutures, and extended apertural neck. USNM 304715

Plate 56, Figure 4a. View opposite that of Plate 55, Figure 4. USNM 304716

Plate 56, Figure 4b. Apertural view showing subtriangular outline, sharply acuminate chambers, moderately strong carinae on final two chambers, and ovate aperture with simple tooth. USNM 304717

Distribution: Common to abundant in Miliolid predominance facies; widespread in low numbers on northern and western shelves and in bays and estuaries. Illustrated specimens from eastern Campeche Shelf.

QUINQUELOCULINA COLLUMNOSA Cushman forma *EXCAVATA*

forma of *Miliolina cuvieriana* Heron-Allen and Earland (not d'Orbigny), 1915, Trans. Zool. Soc. London, vol. 20, pt. 17, p. 571, pl. 4, figs. 33-36.
forma of *Quinqueloculina collumnosa* Cushman, 1922, Carnegie Inst. Wash., Publ. 311, p. 65, pl. 10, fig. 10; Bock, 1971, Miami Geol. Soc., Mem. 1, p. 18, pl. 5, figs. 9-11.

Plate 63, Figure 3. Lateral view showing elliptical outline, four of final five chambers, pock-marked surfaces of chambers, high peripheral carinae, and long apertural neck. USNM 304718

Plate 64, Figure 3a. Lateral view showing elongate terminal extensions of final chamber and strong peripheral carinae. USNM 304719

Plate 64, Figure 3b. Apertural view showing stellate outline, subcircular aperture with small bifid tooth, and sharp peripheral carinae bordering deeply concave chambers. USNM 304719

Distribution: Rare to common in Miliolid predominance facies and other miliolid-rich facies; not known from siliciclastic substrates of northern and western Gulf shelves; forma *excavata* has more distinctly excavated chambers and stronger carinae than forma *typica*. Illustrated specimens from central Campeche Shelf.

QUINQUELOCULINA COMPTA Cushman

Quinqueloculina compta Cushman, 1947, Cushman Lab. Foram. Res., Contr., vol. 23, pt. 4, p. 87, pl. 19, fig. 2; Andersen, 1971, Dept. Conserv., La. Geol. Survey, Bull. 35, pt. 2, p. 29, pl. 4, figs. 2a-c; 3a-c.

Plate 55, Figure 1. Lateral view showing narrow elliptical outline, four of final five chambers, elongate apertural neck, ropy rugose surface caused by interlaced coarse laths of calcite that form outer walls of test, and thin peripheral carina on each chamber. USNM 304720

Plate 56, Figure 1a. View opposite that of Plate 55, Figure 1. USNM 304721

Plate 56, Figure 1b. Apertural view showing subtriangular outline, circular apertural rim with short apertural tooth, ropy rugose surface, narrow peripheral carinae, and truncate margins of two largest chambers. USNM 304722

Distribution: Common in Miliolid predominance facies but also widespread in low numbers elsewhere on northern and western shelves and in bays and estuaries. Illustrated specimens from eastern Campeche Shelf.

QUINQUELOCULINA CRASSA (Heron-Allen and Earland) forma *SUBCUNEATA*

Miliolina crassa Heron-Allen and Earland (part; not d'Orbigny), 1915, Trans. Zool. Soc. London, vol. 20, pt. 17, p. 572, pl. 42, fig. 41.
Quinqueloculina crassa var. *subcuneata* Cushman, 1921, U.S. Nat. Mus. Bull. 100, vol. 4, p. 423, pl. 89, fig. 4; Bock, 1971, Miami Geol. Soc., Mem. 1, p. 18, pl. 5, figs. 12-14.

Plate 61, Figure 1. Lateral view showing broadly elliptical outline, four of final five chambers, strong costae that begin as single ridges at aboral end of chambers and branch adorally; costae converge tangentially with weak carina on third largest chamber. USNM 304723

Plate 62, Figure 1a. View opposite that of Plate 61, Figure 1. USNM 304724

Plate 62, Figure 1b. Apertural view showing tripartite outline, narrow carinae at peripheral angles of chambers merging with branching costae, ovate aperture surrounded by thin lip, and terminally triangulate tooth. USNM 304725

Distribution: Rare to common in Miliolid predominance facies; absent or rare elsewhere. Illustrated specimens from central Campeche Shelf.

QUINQUELOCULINA HORRIDA Cushman

Quinqueloculina horrida Cushman, 1947, Cushman Lab. Foram. Res., Contr., vol. 23, pt. 4, p. 88, pl. 19, fig. 1; Phleger and Parker, 1951, Geol. Soc. America, Mem. 46, pt. II, p. 7, pl. 3, figs. 18, 19; Bock, 1971, Miami Geol. Soc., Mem. 1, p. 19, pl. 6, figs. 1-3.

Plate 59, Figure 2. Lateral view showing elliptical outline, four of final five chambers, coarsely agglutinated outer surface of test constructed chiefly of biogenic carbonate particles, and terminal aperture. USNM 304726

Plate 60, Figure 2a. View opposite that of Plate 59, figure 2. USNM 304727

Plate 60, Figure 2b. Apertural view showing trilobate outline, poorly sorted coarse biogenic fragments cemented in finer grained matrix, smooth circular apertural lip, and short wedge-shaped tooth. USNM 304728

Distribution: Low numbers in Miliolid predominance facies and other shelf biotopes. Illustrated specimens from east Campeche Shelf.

QUINQUELOCULINA LINNEIANA d'Orbigny forma *COMIS*

Triloculina linneiana d'Orbigny var. *comis* Bandy, 1956, U.S. Geol. Survey, Prof. Paper 274-G, p. 198, pl. 29, fig. 12; Bock, 1971, Miami Geol. Soc., Mem. 1, p. 27, pl. 10, figs. 11, 12; pl. 11, fig. 1

Plate 61, Figure 2. Lateral view showing narrow elliptical outline, four of final five chambers, strong rounded costae that are wider than intervening troughs and generally extend the entire length of chambers, and strongly protruding apertural neck. USNM 304729

Plate 62, Figure 2a. View opposite that of Plate 61, Figure 2. USNM 304730

Plate 62, Figure 2b. Apertural view showing weakly tripartite outline, rounded periphery of chambers, strong rounded costae, and subcircular lipped aperture with bifid tooth. USNM 304731

Distribution: Common in Miliolid predominance facies; rare or absent elsewhere. Illustrated specimens from central Campeche Shelf.

QUINQUELOCULINA PARKERI (Brady) forma *OCCIDENTALIS*

Quinqueloculina parkeri (Brady) var. *occidentalis* Cushman, 1921, Proc. U.S. Nat. Mus., vol. 59, p. 69; Bock, 1971, Miami Geol. Soc., Mem. 1, p. 19, pl. 6, figs. 10-12.

Plate 55, Figure 2. Lateral view showing narrowly ovate outline, four of final five chambers, porous ropy surface, and terminal aperture (one side of protruding neck broken). USNM 304732

Plate 56, Figure 2a. View opposite that of Plate 55, Figure 2. USNM 304733

Plate 56, Figure 2b. Apertural view showing strongly tripartite outline, porous ropy surface, thickening of fused "ropes" along periphery of chambers, and circular lipped aperture with strongly bifid tooth. USNM 304734

Distribution: Rare to common in Miliolid predominance facies; absent or rare elsewhere. Illustrated specimens from central Campeche Shelf.

QUINQUELOCULINA TIPSWORDI Andersen

Quinqueloculina tipswordi Andersen, 1961, Dept. Conserv., La. Geol. Survey, Bull. 35, pt. 2, p. 31, pl. 5, figs. 2a-c.

Plate 63, Figure 4. Lateral view showing narrowly ovate outline, four of final five inflated chambers, roughly pitted surface, carina on final chamber, strong basal overhang of final chamber, strongly lipped terminal aperture, short neck, and protruding tooth. USNM 304735

Plate 64, Figure 4a. View opposite that of Plate 63, Figure 4. USNM 304736

Plate 64, Figure 4b. Apertural view showing subdued tripartite outline, roughly pitted surface, bicarinate peripherally concave final two chambers, and crescentic lipped aperture with strongly bifid tooth. USNM 304737

Distribution: Rare in Miliolid predominance facies; rare or absent elsewhere. Illustrated specimens from east Campeche Shelf.

Genus *ROSALINA* d'Orbigny 1826

ROSALINA BAHAMAENSIS Todd and Low

Rosalina candeiana d'Orbigny, 1839, Foram. de Cuba, vol. 8, p. 97, pl. 4, figs. 2-4.
Discorbis candeiana d'Orbigny, Phleger and Parker, 1951, Geol. Soc. America, Mem. 46, pt. II, p. 20, pl. 10, figs. 3a, b.
Rosalina suezensis (Said), Parker, 1954, Mus. Comp. Zool., Harvard, Bull., vol. 111, no. 10, p. 525, pl. 8, figs. 21, 26, 27.
Rosalina bahamaensis Todd and Low, 1971, U.S. Geol. Survey, Prof. Paper 683-C, p. C14, pl. 3, fig. 2.

Plate 41, Figure 3. Umbilical view showing lobate outline, slightly evolute final whorl, coarsely perforate surface, deeply depressed curved radiating sutures, wide deep umbilicus, and narrow imperforate apertural flaps. USNM 304738

Plate 42, Figure 3a. Spiral view showing lobate outline, coarsely perforate surface, imperforate central mound and inner edges of chambers, and depressed curved sutures. USNM 304739

Plate 42, Figure 3b. Edge view showing lateral compression of test, unequal biconvexity, perforate walls on umbilical side and outer edges of chambers on spiral side, imperforate face of final chamber above aperture, and low-lipped aperture reaching periphery. USNM 304740

Distribution: Common in *Rosalina*, *Rosalina-Hanzawaia*, *Rosalina-Asterigerina*, and *Islandiella-Rosalina* predominance facies of West Florida Shelf, widespread in moderate numbers elsewhere on shelf; low numbers on lower slope may have been displaced from shelf. Illustrated specimens from Texas-Louisiana Shelf edge.

ROSALINA CONCINNA (Brady)

Discorbina concinna H. B. Brady, 1884, Rept. Voy. *Challenger*, Zool., vol. 9, p. 646, pl. 90, figs. 7, 8.
Rosalina cf. *concinna* (H. B. Brady), Parker, 1954, Mus. Comp. Zool., Harvard, Bull., vol. 111, no. 10, p. 524, pl. 8, figs. 17, 18.

Plate 41, Figure 4. Umbilical view showing subcircular outline, four crescentic chambers of final whorl, coarsely perforate outer parts of chambers, imperforate umbilical parts of chambers, deep wide subquadrate umbilicus, slightly curved depressed sutures, final chamber composing nearly one-half final whorl, and aperture opening under broken umbilical flap. USNM 304741

Plate 42, Figure 4a. Spiral view showing subcircular outline, crescentic chambers, coarsely perforate surface, open pores concentrated along sutures, and thickened curved sutures. USNM 304742

Plate 42, Figure 4b. Edge view showing high spire, flattened umbilical side, thickened imperforate periphery, and coarse open pores along thickened sutures. USNM 304743

Distribution: Especially abundant in *Rosalina*, *Rosalina-Hanzawaia*, *Islandiella-Rosalina*, and *Rosalina-Asterigerina* predominance facies on West Florida Shelf; sparsely reported elsewhere but may have been included by some authors under name *Neoconorbina orbicularis* (Terquem), closely similar form. Illustrated specimens from West Florida Shelf.

ROSALINA FLORIDENSIS (Cushman)

Discorbis bertheloti (d'Orbigny) var. *floridensis* Cushman, 1930, Jour. Paleont., vol. 40, no. 4, p. 364, pl. 33, figs. 13a-c.
Rosalina floridensis (Cushman), Parker, 1954, Mus. Comp. Zool., Harvard, Bull., vol. 111, no. 10, p. 525, pl. 8, figs. 28, 29.

Plate 41, Figure 2. Umbilical view showing subcircular outline, greatly embracing chambers, deep stellate umbilicus, densely perforate outer portions of chambers, imperforate inner portions of chambers extending as triangular flaps into the umbilicus, imperforate peripheral carina, and aperture opening under umbilical flap and confined to umbilicus. USNM 304744

Plate 42, Figure 2a. Spiral view showing subcircular outline, densely perforate walls, narrow curved imperforate sutures, and narrow imperforate peripheral carina. USNM 304745

Plate 42, Figure 2b. Edge view showing planoconvex outline, densely perforate spiral chambers separated by imperforate sutures, thick imperforate keel, and aperture opening into umbilicus. USNM 304746

Distribution: Widespread in moderate numbers along inner to outer shelf, especially in *Rosalina*, *Rosalina-Hanzawaia*, and *Islandiella-Rosalina* predominance facies of northern West Florida Shelf. Illustrated specimens from Texas Shelf.

ROSALINA SUBARAUCANA (Cushman)

Discorbis subaraucana Cushman, 1922, Carnegie Inst. Washington, Publ. 311, vol. 17, p. 41, pl. 7, figs. 1, 2.
Rosalina subaraucana (Cushman), Wantland, 1975, Amer. Assoc. Petrol. Geol., Stud. Geol. No. 2, p. 395, fig. 60p.

Plate 41, Figure 1. Umbilical view showing faintly lobate outline, six chambers of final whorl, coarsely and densely perforate outer portions of walls, broad imperforate patches surrounding narrow deep umbilicus, broad imperforate sutures between early chambers, thick imperforate peripheral band in first three visible chambers, broad spatulate flaps over aperture extending well into umbilicus and overlapping preceding flaps, and aperture that does not reach periphery. USNM 304747

Plate 42, Figure 1a. Spiral view showing slightly lobate periphery, coarsely and densely perforate walls, and wide imperforate curved sutures. USNM 304748

Plate 42, Figure 1b. Edge view showing thick imperforate peripheral band in early coil, imperforate sutures separating coarsely perforate chambers on spiral side, and low arched aperture with narrow lipped flap; aperture does not reach periphery. USNM 304749

Distribution: Moderately abundant in *Rosalina* predominance facies and other carbonate or mixed siliciclastic-carbonate biotopes; scattered occurrences elsewhere on shelf. Illustrated specimens from Texas Shelf.

Genus *SACCAMMINA* M. Sars, 1869

SACCAMMINA DIFFLUGIFORMIS (H. B. Brady) forma *TYPICA*

Reophax difflugiformis H. B. Brady, 1879, Quart. Jour. Micr. Sci., vol. 19, p. 51, pl. 4, figs. 3a, b.

Proteonina difflugiformis (H. B. Brady), Phleger and Parker, 1951, Geol. Soc. America, Mem. 46, pt. II, p. 2, pl. 1, fig. 5 (fig. 6 is forma *atlantica* Cushman).

Plate 49, Figure 1. Lateral view showing guttulate outline and coarse terrigenous mineral grains set loosely in finer grained matrix. USNM 304750

Plate 50, Figure 1a. Apertural view showing fine-grained apertural neck and coarser grained outer wall surface. USNM 304750

Plate 50, Figure 1b. Apertural enlargement. USNM 304750

Distribution: Copredominant in *Saccammina-Ammobaculites* predominance facies of Texas-Louisiana Shelf; widespread in moderate numbers in other shelf facies; low numbers on slope, rise, and Sigsbee Plain. Illustrated specimen from Texas Shelf.

Genus *SACCORHIZA* Eimer and Fickert, 1899

SACCORHIZA RAMOSA (H. B. Brady)

Hyperammina ramosa H. B. Brady, 1879, Quart. Jour. Micr. Sci., new ser., vol. 19, p. 33, pl. 3, figs. 14, 15.

Plate 15, Figure 4. Lateral view showing branching tubular construction of test; outer wall formed of poorly sorted large terrigenous mineral grains and biogenic particles set in finer grained matrix. USNM 304751

Plate 16, Figure 4a. Enlargement of branch tip showing coarse grains on inner and outer wall surfaces and separated by layer of fine-grained detritus; some large fragments extend through entire thickness of wall. USNM 304751

Plate 16, Figure 4b. Enlargement of surface showing large conchoidally fractured mineral grains and fractured sponge spicule set in matrix of finer particles. USNM 304751

Distribution: Common to abundant in agglutinant *Glomospira* predominance facies of slope, rise, and Sigsbee Plain; usually occurs as broken fragments, making it difficult to assess original abundance of unbroken specimens. Illustrated specimen from Sigsbee Plain.

Genus *SIPHOTROCHAMMINA* Saunders, 1957

SIPHOTROCHAMMINA SQUAMATA (Jones and Parker)

Trochammina squamata (Jones and Parker), 1860, Geol. Soc. London, Quart. Jour., vol. 16, p. 304; LeRoy and Hodgkinson, 1975, Micropaleont., vol. 21, p. 438, pl. 5, figs. 4, 5.

Plate 13, Figure 2. Umbilical view showing lobate outline, flattened chamber faces, large deep quadrate umbilicus, depressed sigmoid sutures, low arched aperture at edge of umbilicus, and wall constructed of poorly sorted angular terrigenous mineral grains arranged with flat faces parallel to chamber surfaces and cemented into finer grained matrix. USNM 304752

Plate 14, Figure 2a. Enlargement of aperture showing position at edge of umbilicus, arched shape, and small mineral grains forming rim of aperture. USNM 304752

Plate 14, Figure 2b. Edge view showing concavoconvex shape, flattened chambers of umbilical side, and rugose surface of spiral side contrasted with smoother surface of umbilical side. USNM 304752

Plate 14, Figure 2c. Spiral view showing lobate outline, inflated chambers, distinct depressed sutures, and poorly sorted larger mineral grains set in finer matrix; surface more rugose than umbilical side because flat faces of grains do not parallel chamber surfaces. USNM 304752

Distribution: Common to rare in scattered locations from inner shelf to Sigsbee Plain; often in agglutinant *Glomospira* predominance facies. Illustrated specimen from Louisiana Slope.

Genus *SPIROLOCULINA* d'Orbigny, 1826

SPIROLOCULINA ANTILLARUM d'Orbigny forma *CRENULATA*

forma of *Spiroloculina antillarum* d'Orbigny, 1839, Foram. de Cuba, vol. 8, p. 166, pl. 9, figs. 3, 4; Bock, 1971, Miami Geol. Soc., Mem. 1, p. 13, pl. 3, fig. 7.
?*Spiroloculina grata* Terquem, 1878, Geol. Soc. France, Mém. sér. 3, tome 1, no. 3, p. 55, pl. 5, figs. 14a-b, 15a-b.

Plate 53, Figure 3. Lateral view showing narrowly ovate outline, coiled alternating chamber arrangement, crenulate longitudinal costae, protruding apertural neck, and basal overhang of final chamber. USNM 304753

Plate 54, Figure 3a. Edge view showing planar coiling of test, crenulate longitudinal costae, and protruding apertural neck. USNM 304754

Plate 54, Figure 3b. Apertural view showing crenulate costae with wide intervening troughs, subcircular aperture surrounded by thin lip, and short wedge-shaped apertural tooth, USNM 304755

Distribution: Rare to common in Miliolid predominance facies and in low numbers in other carbonate-rich or mixed siliciclastic-carbonate biotopes; differs from *A. antillarum* forma *typica* in possessing crenulate longitudinal costae. Illustrated specimens from east Campeche Shelf.

SPIROLOCULINA SOLDANII Fornasini forma *DENTATA*

Spiroloculina dentata Cushman and Todd, 1944, Cushman Lab. Foram. Res., Spec. Publ. 11, p. 71, pl. 9, figs. 33, 34.

Spiroloculina cf. *S. dentata* Cushman and Todd, Andersen, 1961, Dept. Conserv., La. Geol. Survey, Bull. 35, pt. 2, p. 35, pl. 5, figs. 8a, b.

Plate 53, Figure 1. Lateral view showing broadly ovate shape, evolute coil of alternating chambers, concave center of test, moderately rougose surface texture, and terminal toothed aperture. USNM 304756

Plate 54, Figure 1a. Edge view showing flattened periphery, lateral compression of test, and rectangular aperture with tooth. USNM 304757

Plate 54, Figure 1b. Apertural view showing flattened periphery, biconcavity of test, sharp peripheral angles of final two chambers, subrectangular aperture, and elongate bifid tooth. USNM 304758

Distribution: Rare to common in Miliolid predominance facies and in other carbonate-rich biotopes of shelf. Illustrated specimens from central Campeche Shelf.

SPIROLOCULINA SOLDANII Fornasini forma *TYPICA*

Spiroloculina soldanii Fornasini, 1886, Boll. Soc. Geol. Ital., vol. 5, p. 25; Bock, 1971, Miami Geol. Soc., Mem. 1, p. 15, pl. 4, fig. 1.

Plate 53, Figure 2. Lateral view showing broadly ovate outline, evolute coil of alternating chambers, depressed center of coil, slightly rugose surface texture, depressed sutures, and terminal aperture. USNM 304759

Plate 54, Figure 2a. Edge view showing lateral compression of test, peripheral carinae of penultimate chamber, and rectangular aperture with bifid tooth. USNM 304760

Plate 54, Figure 2b. Apertural view showing biconcavity of test, convexly truncate periphery of final two chambers, carinate chamber margins, slightly rugose surface texture, roundly rectangular aperture, and strongly bifid tooth. USNM 304761

Distribution: Rare to common in Miliolid predominance facies and other carbonate-rich biotopies of shelf. Illustrated specimens from east Campeche Shelf.

Genus *TRILOCULINA* d'Orbigny, 1826

TRILOCULINA TRICARINATA d'Orbigny

Triloculina tricarinata d'Orbigny, 1826, Ann. Sci. Nat., vol. 7, p. 299, no. 7; Bock, 1971, Miami Geol. Soc., Mem. 1, p. 28, pl. 12, figs. 1, 2.

Plate 57, Figure 3. Lateral view showing broadly ovate outline, three chambers of final whorl, smooth flattened surfaces of chambers, sharp carina of penultimate chamber, and toothed terminal aperture. USNM 304762

Plate 58, Figure 3a. View opposite that of Plate 57, Figure 3. USNM 304762

Plate 58, Figure 3b. Apertural view showing subtriangular outline, strong single carinae at peripheral angles, smooth slightly convex chambers, ovate lipped aperture, and short bifid tooth. USNM 30462

Distribution: Rare to common in Miliolid predominance facies and other carbonate-rich biotopes of shelf. Illustrated specimen from central Campeche Shelf.

TRILOCULINA TRIGONULA (Lamarck)

Miliola trigonula Lamarck, 1804, Mus. Nat. Hist. Paris, Ann., vol. 5, no. 3, p. 351.
Triloculina trigonula (Lamarck), Bock, 1971, Miami Geol. Soc., Mem. 1, p. 28, pl. 12, figs. 3, 4.

Plate 57, Figure 2. Lateral view showing broadly ovate outline, final three chambers, smooth (but bored) surface, narrow depressed sutures, terminal aperture, and strongly protruding tooth. USNM 304763

Plate 58, Figure 2a. Edge view showing broadly ovate outline, final three chambers, depressed sutures, and protruding tooth. USNM 304764

Plate 58, Figure 2b. Apertural view showing rounded triangular outline, flattened chambers, smooth surface, and ovate aperture with strongly protruding bifid tooth. USNM 304763

Distribution: Common in Miliolid predominance facies and other carbonate-rich biotopes of shelf. Illustrated specimens from central Campeche Shelf.

TRILOCULINA VARIOLATA d'Orbigny

Quinqueloculina variolata d'Orbigny, 1826, Ann. Sci. Nat., p. 302, no. 26; Planches inedite, pl. 5, fig. 5; Le Calvez, 1977, Cah. Micropaleont., no. 1, p. 57, pl. 8, figs. 1-7.
Triloculina bicarinata d'Orbigny, 1839, Foram. de Cuba, vol. 8, p. 158, pl. 10, figs. 18-20; Bock, 1971, Miami Geol. Soc., Mem. 1, p. 25, pl. 9, figs. 12, 13; pl. 10, fig. 1.
Triloculina carinata d'Orbigny, 1839, Foram. de Cuba, vol. 8, p. 189, pl. 11, figs. 15-17.
Quinqueloculina antillarum d'Orbigny, 1839, Foram. de Cuba, vol. 8, p. 194, pl. 12, figs. 4-6.

Plate 57, Figure 1. Lateral view showing broadly ovate outline, three final chambers, ropy reticulate surface with fusiform pitting which is subdued at test extremities, and terminal aperture. USNM 304765

Plate 58, Figure 1a. View opposite that of Plate 57, Figure 1. USNM 304766

Plate 58, Figure 1b. Apertural view showing reticulate and pitted surface, bicarinate final chamber, unicarinate penultimate chamber, elongate narrow lipped aperture, and long bifid tooth. USNM 304767

Distribution: Rare to common in Miliolid predominance facies; rare in other mixed siliciclastic-carbonate biotopes of shelf. Illustrated specimens from central Campeche Shelf.

TRILOCULINA sp.

Plate 57, Figure 4. Lateral view showing broadly ovate outline, three chambers of final whorl, subdued widely spaced costae, depressed sutures, and terminal aperture. USNM 304768

Plate 58, Figure 4a. View opposite that of Plate 57, Figure 4. USNM 304769

Plate 58, Figure 4b. Apertural view showing rounded triangular outline, flattened chambers, strong costae on penultimate chamber, subdued costae on other two chambers, subquadrate lipped aperture, and strong bifid tooth. USNM 304770

Distribution: Low numbers in Miliolid predominance facies; rare in other carbonate-rich biotopes of shelf. Illustrated specimens from central Campeche Shelf.

Genus *TROCHAMMINA* Parker and Jones, 1859

TROCHAMMINA GLOBIGERINIFORMIS (Parker and Jones)

Lituola nautiloidea var. *globigeriniformis* Parker and Jones, 1865, Trans. Roy. Soc. London, vol. 155, p. 407, pl. 15, figs. 46, 47.
Haplophragmium globigeriniformis (Parker and Jones), Brady, 1884, Rept. Voy. *Challenger*, p. 312, pl. 35, figs. 10a-c.

Plate 13, Figure 1. Umbilical view showing distinctly lobate subcircular outline, four inflated chambers of final whorl with narrow arched aperture opening into narrow umbilicus, chamber walls constructed of well-sorted, coarse, flattened terrigenous mineral grains set in finer grained matrix, and finer grained matrix forming narrow bridge across aperture. USNM 304771

Plate 14, Figure 1a. Edge view showing biconvex outline, globular chambers, umbilical aperture, and flat-grained wall structure. USNM 304772

Plate 14, Figure 1b. Spiral side showing lobate subcircular outline, few inflated chambers, depressed sutures, and well-sorted flat grains of surface layer in finer grained matrix. USNM 304773

Plate 14, Figure 1c. Enlargement of aperture showing fine-grained structure of apertural bridge. USNM 304771

Distribution: Common in agglutinant *Glomospira* predominance facies and in other locations on lower slope, rise, and Sigsbee Plain. Illustrated specimens from Louisiana Slope.

TROCHAMMINA(?) sp.

Trochammina quadriloba Höglund, Parker, 1954, Mus. Comp. Zool., Harvard, Bull., vol. 111, no. 10, p. 503, pl. 5, figs. 14, 15 (not Höglund, 1947).

Plate 13, Figure 3. Umbilical view showing somewhat lobate slightly subcircular outline, four chambers of final whorl, straight depressed radiating sutures, deep narrow umbilicus, outer wall composed of well-sorted large terrigenous mineral grains loosely set in finer grained matrix, and large spatulate flap of fine grains extending over aperture into umbilicus; first two chambers of final whorl broken near umbilicus. USNM 304774

Plate 14, Figure 3a. Edge view showing low spire, rounded periphery, nearly planoconvex outline, deep umbilicus covered partly by protruding fine-grained apertural lip, and well-sorted large mineral grains loosely set in finer matrix of outer wall. USNM 304775

Plate 14, Figure 3b. Spiral view showing subspherical lobate outline, proloculus and succeeding chambers of low spire, narrow depressed sutures, and well-sorted large mineral grains loosely set in finer matrix to form outer layer of chamber walls. USNM 304776

Distribution: Common on middle and outer shelf of northeastern Gulf; low numbers elsewhere down to Sigsbee Plain; often in *Glomospira* predominance facies. Illustrated specimens from Texas-Louisiana Slope.

Genus *UVIGERINA* d'Orbigny, 1826

UVIGERINA FLINTII Cushman

Uvigerina flintii Cushman, 1923, U.S. Nat. Mus., Bull., vol. 104, pt. 4, p. 165, pl. 42, fig. 13; Parker, 1954, Mus. Comp. Zool., Harvard, Bull., vol. 111, pt. 10, p. 520, pl. 8, fig. 2.

Plate 27, Figure 5. Lateral view showing ovate outline, triserial coiling of chambers, elongate spirally costate neck, widely spaced rounded costae crossing depressed sutures between early chambers and often fusing with costae of previous chambers; fine perforations of wall not visible at this magnification. USNM 304777

Plate 28, Figure 5a. Apertural view showing subcircular outline, triserial coiling, laterally truncate final chamber, strong, widely spaced, smoothly rounded costae, circular aperture with tooth-plate, and dentate spiral costa around neck. USNM 304778

Plate 28, Figure 5b. Prolocular view showing slightly rugose proloculus and radiating, strong, smoothly rounded costae. USNM 304779

Distribution: Widespread in moderate numbers on outer shelf and upper slope, especially in *Brizalina*, *Uvigerina*, and *Brizalina-Uvigerina* predominance facies. Illustrated specimens from Texas-Louisiana Shelf edge.

UVIGERINA LAEVIS Goës

Uvigerina auberiana Goës, 1882 (not d'Orbigny, 1839), Kongl. Svensk. Vet. Akad. Handl., vol. 19, no. 4, p. 60, pl. 4, figs. 71-74.
Uvigerina auberiana d'Orbigny var. *laevis* Goës, 1896, Mus. Comp. Zool., Harvard, Bull., vol. 29, p. 51.
Uvigerina laevis Goës, Parker, 1954, Mus. Comp. Zool., Harvard, Bull., vol. 111, no. 10, p. 520, pl. 8, fig. 4.

Plate 27, Figure 4. Lateral view showing narrow lenticular outline, triserial coiling of chambers, strong basal spine, finely hispid surface, markedly looser coiling in final three chambers, and progressively finer hispidity of adult chambers. USNM 304780

Plate 28, Figure 4a. Apertural view showing subcircular outline, densely hispid surface, laterally truncate final chamber, circular aperture surrounded by long hispid neck, everted apertural lip largely broken off, and curved tooth-plate. USNM 304781

Plate 28, Figure 4b. Prolocular view showing subcircular outline and more coarsely hispid surface of early chambers. USNM 304782

Distribution: Common to abundant in shelf edge and upper slope facies; one of three most abundant forms of genus in *Uvigerina* and *Brizalina-Uvigerina* predominance facies; ranges in lower numbers from inner shelf to Sigsbee Plain. Illustrated specimens from Texas-Louisiana Shelf edge.

UVIGERINA PEREGRINA Cushman forma *HISPIDOCOSTATA*

Uvigerina hispidocostata Cushman and Todd, 1945, Cushman Lab. Foram. Res., Spec. Publ. 15, p. 51, pl. 7, figs. 27, 31; Parker, 1954, Mus.

Comp. Zool., Harvard, Bull., vol. 111, no. 10, p. 520, pl. 8, fig. 3.
Uvigerina peregrina dirupta Todd, Pflum and Frerichs, 1976, Jour. Foram. Res., Spec. Publ. No. 14, pl. 8, figs. 4, 5.
Uvigerina spinicostata Cushman and Jarvis, Pflum and Frerichs, 1976, Jour. Foram. Res., Spec. Publ. No. 14, pl. 8, figs. 6, 7.

Plate 27, Figure 2. Lateral view showing broadly ovate shape, triserial coiling of chambers, widely spaced, high, thin costae forming peaked overhangs along chamber bases but rarely crossing sutures, flat-bottomed intervening troughs containing rows of stubby spinules and fine scattered pores, and larger spines on final chamber and on apertural neck. USNM 304783

Plate 28, Figure 2a. Apertural view showing subcircular outline, widely spaced, high, sharp costae, flat-bottomed intervening troughs with stubby spinules, hispid apertural neck, and circular aperture with serrate tooth-plate. USNM 304784

Plate 28, Figure 2b. Prolocular view showing subcircular outline, widely spaced, high, sharp costae, and flat-bottomed intervening troughs with stubby spinules. USNM 304785

Distribution: Moderately abundant in shelf-edge and upper and middle slope facies around Gulf; ranges in lower numbers from middle shelf to Sigsbee Plain. Illustrated specimens from Texas Shelf.

UVIGERINA PEREGRINA Cushman forma *PARVULA*

Uvigerina peregrina Cushman var. *parvula* Cushman, 1923, U.S. Nat. Mus., Bull. 104, pt. 4, p. 168, pl. 42, fig. 11.
Uvigerina parvula Cushman, Parker, 1954, Mus. Comp. Zool., Harvard, Bull., vol. 111, no. 10, p. 521, pl. 8, fig. 6.

Plate 27, Figure 3. Lateral view showing fusiform shape, triserial coiling of chambers, low, sharp, widely spaced costae, some costae with slightly serrate outer edges, costae usually forming peaked overhang at chamber bases and seldom crossing sutures, intervening troughs containing a few scattered short spinules, and final chamber more densely spinulose with elongate apertural neck; fine perforations not visible at this magnification. USNM 304786

Plate 28, Figure 3a. Apertural view showing subcircular outline, strong, moderately serrate costae, spinulose final chamber strongly truncate at one side, circular aperture surrounded by spinulose neck, curved serrate tooth-plate, and scattered spinules in intercostal troughs. USNM 304787

Plate 28, Figure 3b. Prolocular view showing subcircular outline and widely spaced costae with slightly serrate edges. USNM 304788

Distribution: Most abundant and widespread form of *Uvigerina* in Gulf; especially abundant at shelf edge and on upper slope; one of three most abundant forms of genus in *Uvigerina* and *Brizalina-Uvigerina* predominance facies; ranges in lower numbers from bays and estuaries to Sigsbee Plain. Illustrated specimens from Texas-Louisiana Shelf edge.

UVIGERINA PEREGRINA Cushman forma *TYPICA*

Uvigerina peregrina Cushman, 1923, U.S. Nat. Mus., Bull. 104, pt. 4, p. 166, pl. 42, figs. 7-10; Parker, 1954, Mus. Comp. Zool., Harvard, Bull., vol. 111, no. 10, p. 521, pl. 8, fig. 5.
Uvigerina peregrina peregrina Cushman, Pflum and Frerichs, 1976, Jour. Foram. Res., Spec. Publ. No. 14, pl. 8, figs. 2, 3.
Uvigerina peregrina mediterranea Hofker, Pflum and Frerichs, 1976, Jour. Foram. Res., Spec. Publ. No. 14, pl. 8, fig. 1.

Plate 27, Figure 1. Lateral view showing narrowly ovate outline, triserial coiling of chambers, strong, high costae rarely crossing sutures, costae coarsely serrate and even spinose along outer margins in early chambers; final chamber without costae but with coarse, densely scattered spines extending up onto apertural neck; fine perforations not visible at this magnification. USNM 304789

Plate 28, Figure 1a. Apertural view showing subcircular outline, strong costae with serrate and spinose edges, spinose final chamber, and circular aperture with tooth-plate. USNM 304790

Plate 28, Figure 1b. Prolocular view showing subcircular outline, strong serrate costae, and long prolocular spines. USNM 304791

Distribution: Widespread in moderate to large numbers in outer shelf and upper slope facies; one of three most abundant forms of genus in *Uvigerina* and *Brizalina-Uvigerina* predominance facies; ranges in lower numbers from inner shelf to Sigsbee Plain. Illustrated specimens from East Mexico Slope.

Genus *WIESNERELLA* Cushman, 1933

WIESNERELLA AURICULATA (Egger)

Planispirina auriculata Egger, 1893, Abhandl. K. Bayer Akad. Wiss, München, vol. 18, pt. 2, p. 245, pl. 3, figs. 13-15.
Wiesnerella auriculata (Egger), Parker, 1954, Mus. Comp. Zool., Harvard, Bull., vol. 111, no. 10, p. 501, pl. 5, fig. 13.

Plate 53, Figure 4. Lateral view showing broadly ovate outline, alternating spiral chamber arrangement, finely striate surface, and flaring subcircular apertural lip. USNM 304792

Plate 54, Figure 4a. Edge view showing strongly flattened test, narrow peripheral carina, and flaring lip of terminal aperture. USNM 304793

Plate 54, Figure 4b. View opposite that of Plate 54, Figure 4, showing ovate outline, finely striate surface and flaring apertural lip. USNM 304794

Distribution: Common to rare in inner to middle shelf facies, especially on carbonate-rich substrates. Illustrated specimens from east Campeche Shelf.

SCANNING ELECTRON PHOTOMICROGRAPHS

To paraphrase Ager (1963), the first principle of practical ecology or paleoecology is that one should go to the field and look at the organisms, or their remains, carefully. Prior to the wide use of the scanning electron microscope, this "look" at benthic foraminifera met with only moderate success. Many of the morphologic features are dimly visible using optical microscopy. Elton (1927) and Whittington (1964) have emphasized that progress in ecology and paleoecology depends on the accurate identification of whatever taxa are studied. For this task, the foraminiferal specialist relies primarily on the morphology of the foraminiferal test. The photomicrographs on the following 64 plates illustrate the detailed external morphology of 138 of the abundant or ecologically diagnostic species belonging to the predominant benthic genera in the Gulf. Morphologic descriptions of each species illustrated are provided in the preceding section. The scanning electron photomicrographs should promote consistent identification of those species included and of the genera to which they are assigned. Furthermore, they demonstrate how valuable the scanning electron microscope can be for taxonomic studies of exterior test morphology when care is taken to produce good photomicrographs.

The photomicrographs are grouped by genus according to their bathymetric distribution. Plates with low numbers illustrate species whose genera form deep-water predominance facies. Plates with high numbers illustrate species whose genera form shallow-water predominance facies. For each pair of facing plates, the left-hand plate shows enlarged lateral views of species belonging to genera whose names appear at the bottom of the plate; the right-hand plate shows reduced alternate views of the same species and indicates the facies in which these species are most abundant.

Illustrated specimens were mounted on aluminum stubs with double-coated Scotch tape, coated with a film of gold-paladium, and photographed with Polaroid type 55 black-and-white film using an ETEC Autoscan U-1 scanning electron microscope (the citation of trade names is for descriptive purposes only and does not constitute endorsement of the products by the U.S. Geological Survey).

The bar scales on each plate represent a length of 50 μm. Illustrated specimens are deposited in the U.S. National Museum, Washington, D.C. (USNM numbers 304427 to 304794).

PLATE 1

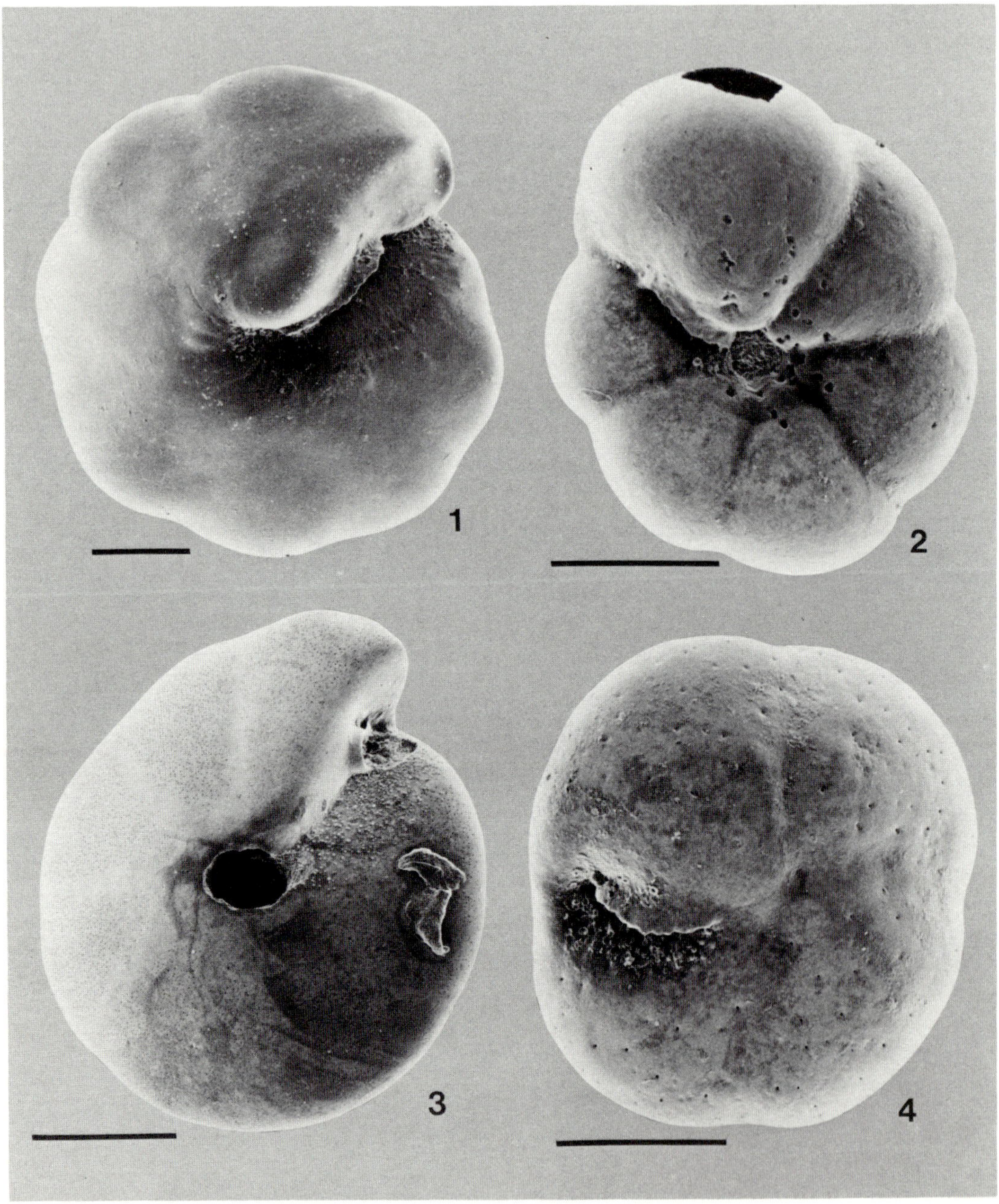

EPONIDES

PLATE 2

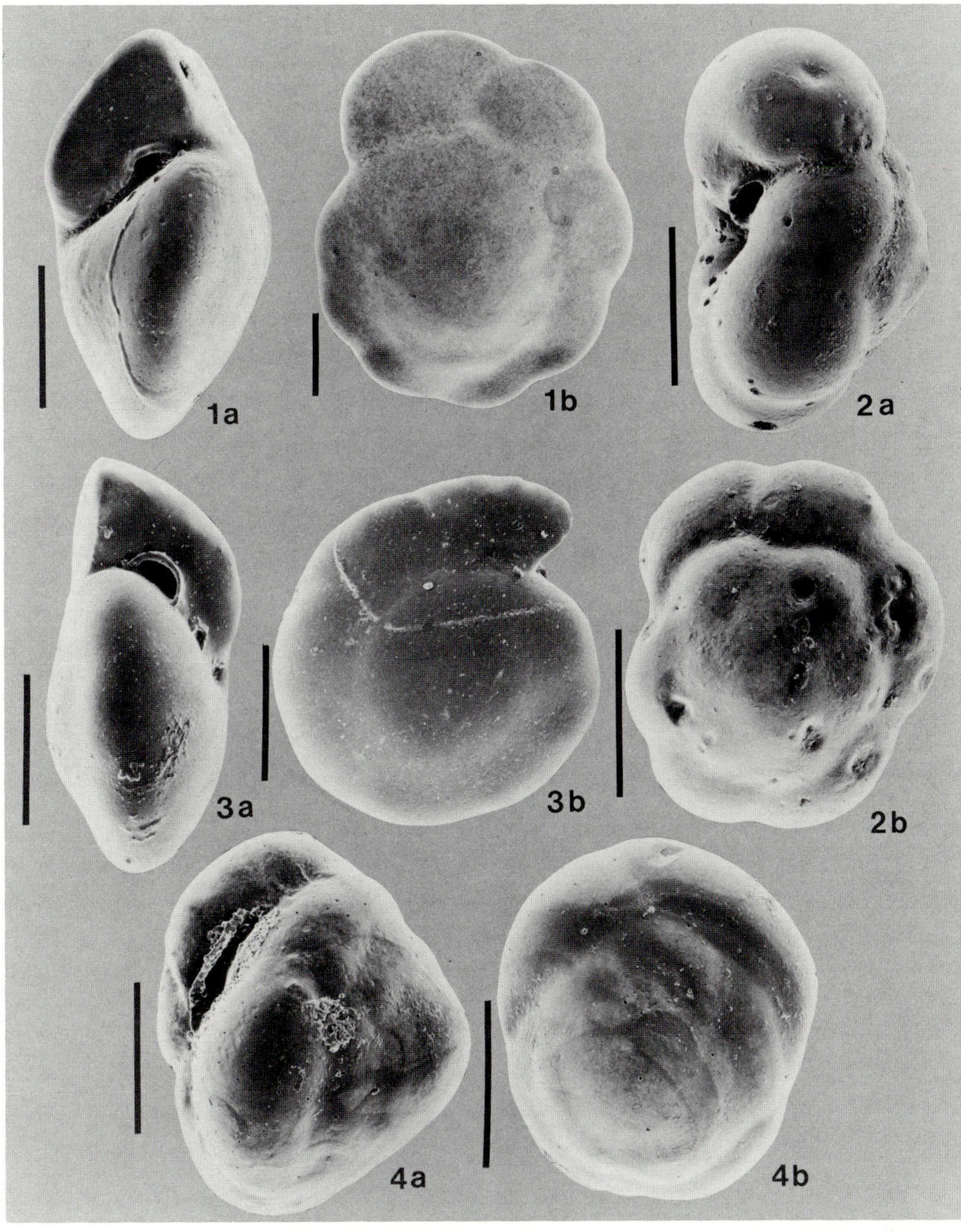

ABYSSAL FACIES

PLATE 3

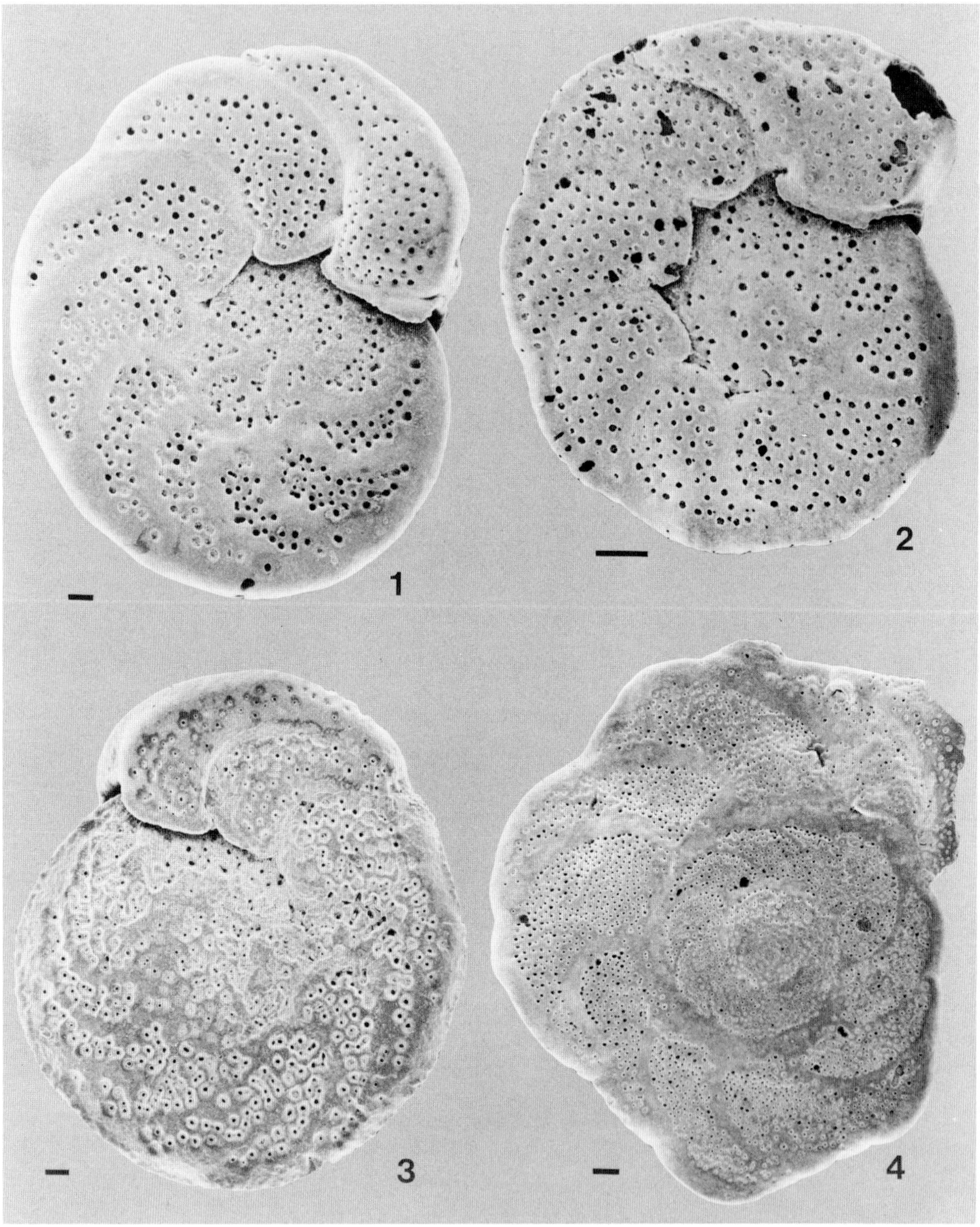

CIBICIDES

PLATE 4

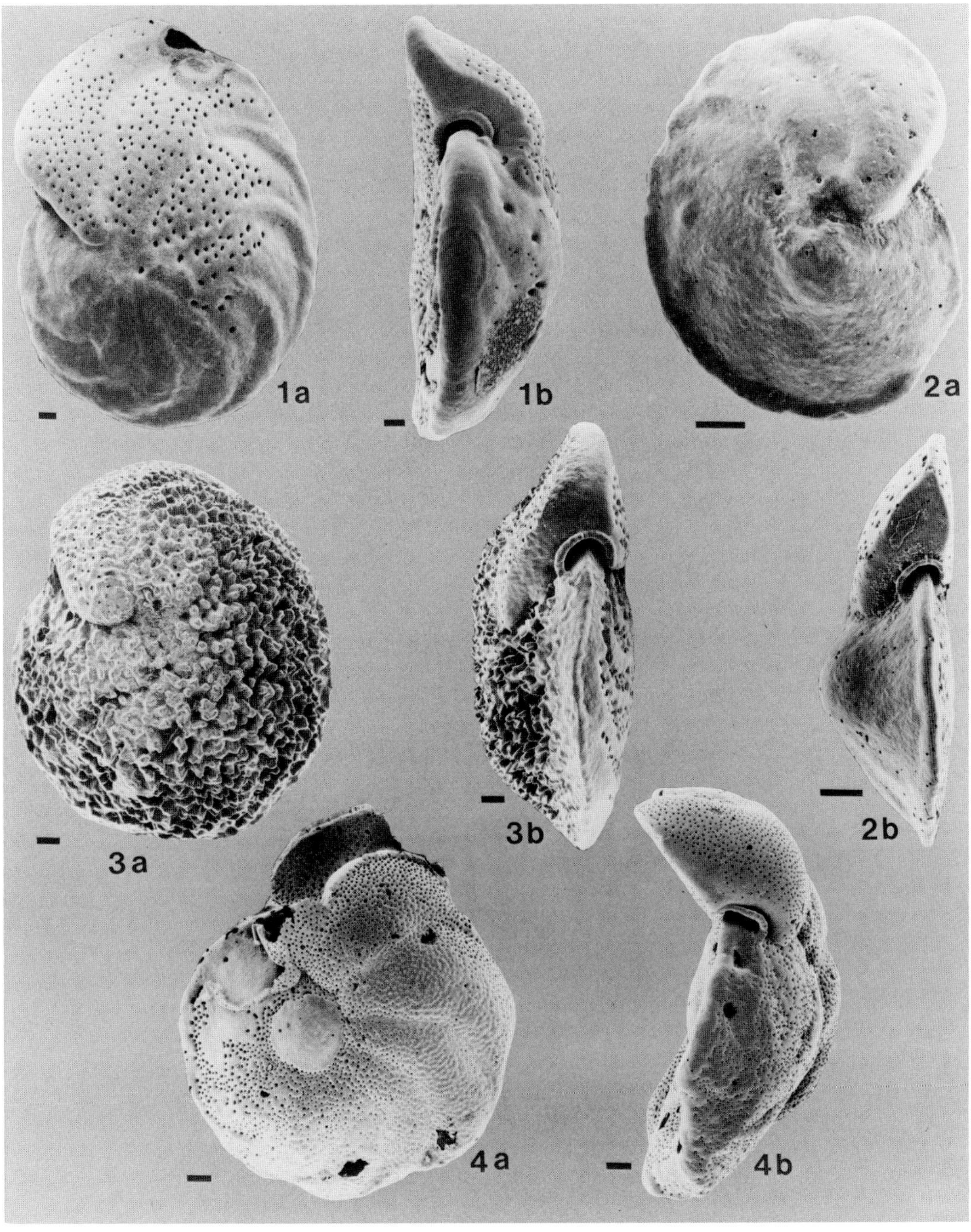

SHELF, SLOPE, AND ABYSSAL FACIES

PLATE 5

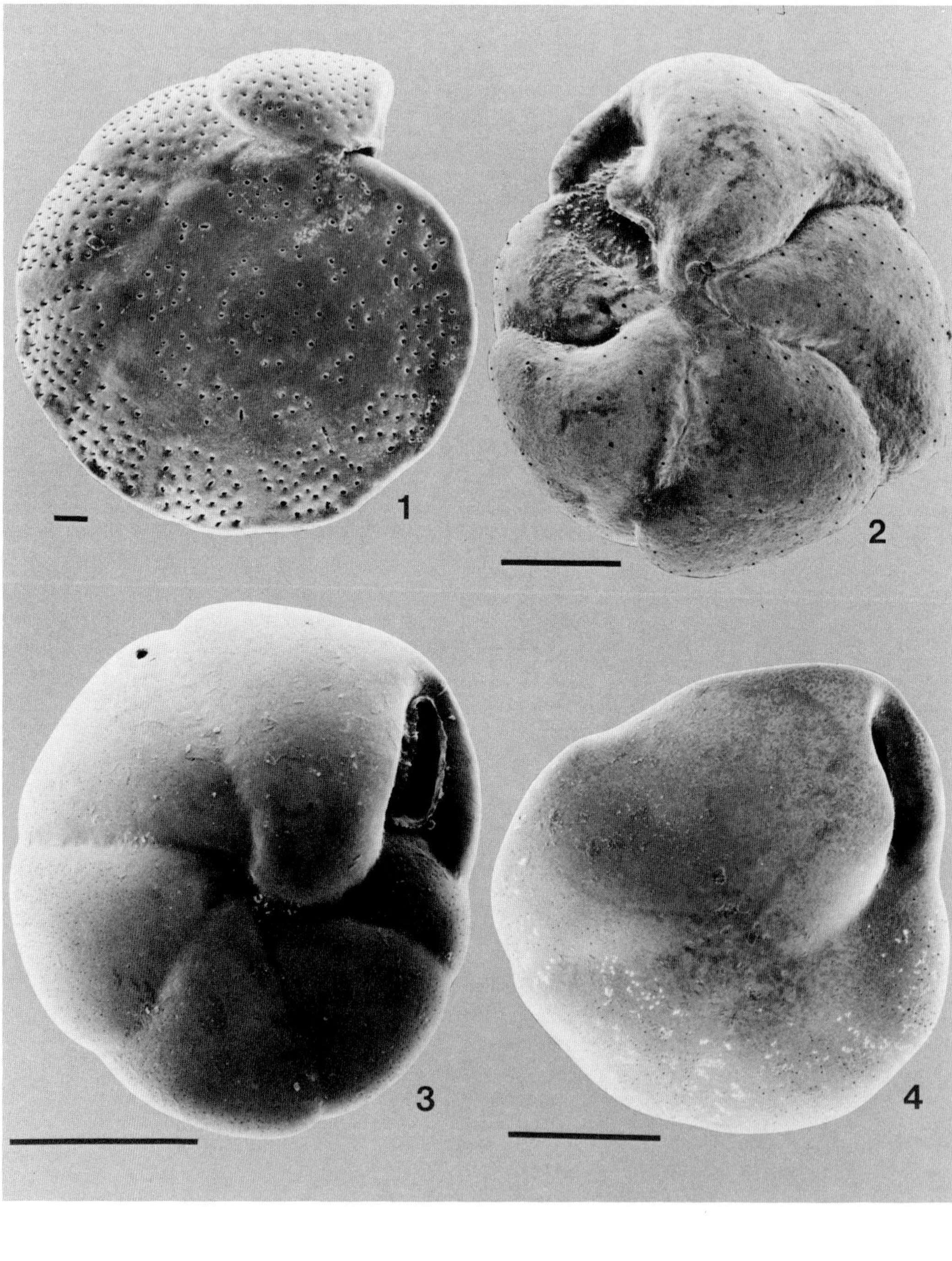

CIBICIDOIDES, NUTTALLIDES, EPISTOMINELLA

PLATE 6

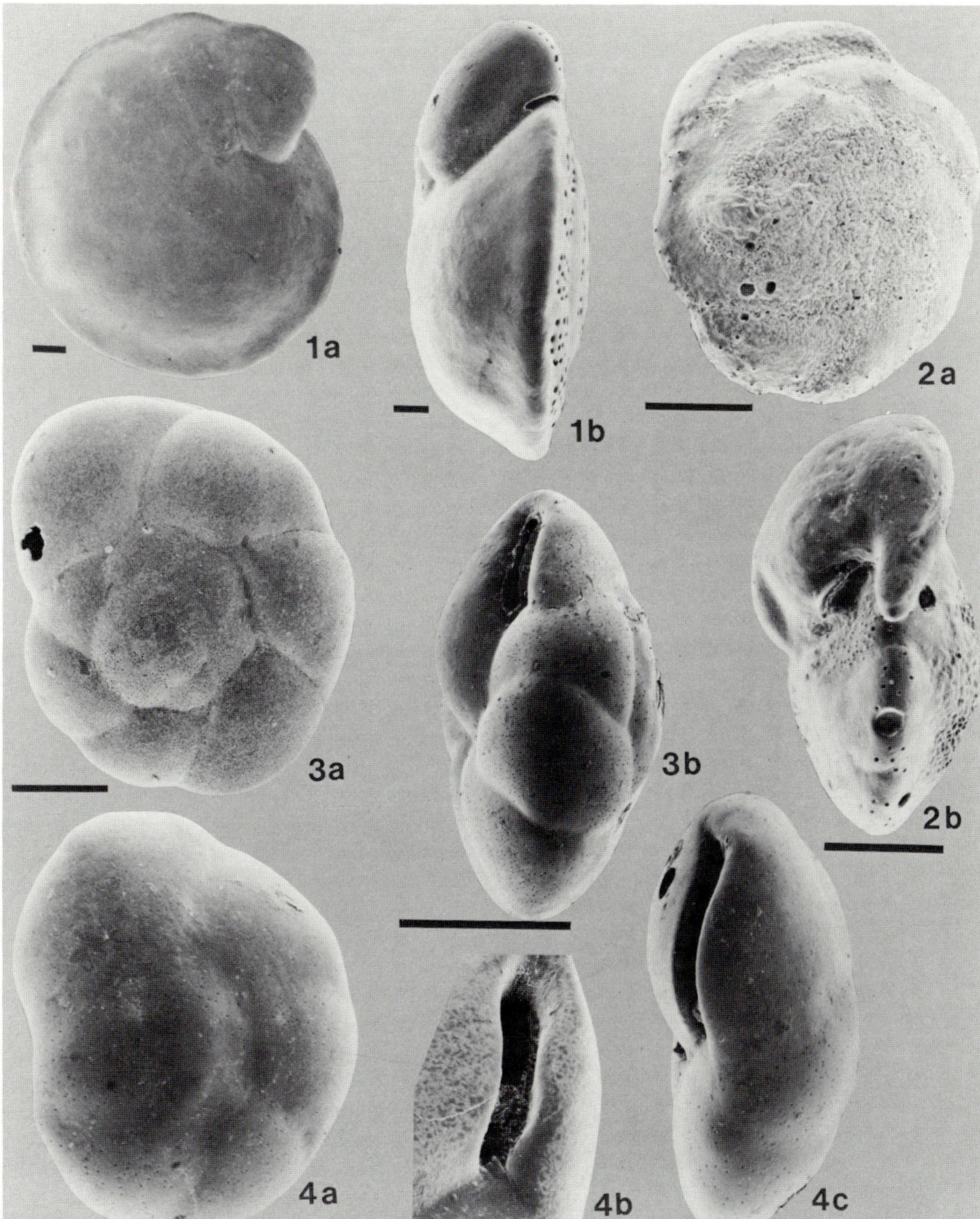

SHELF, SLOPE, AND ABYSSAL FACIES

PLATE 7

AMMODISCUS, AMMOBACULITES, GLOMOSPIRA

PLATE 8

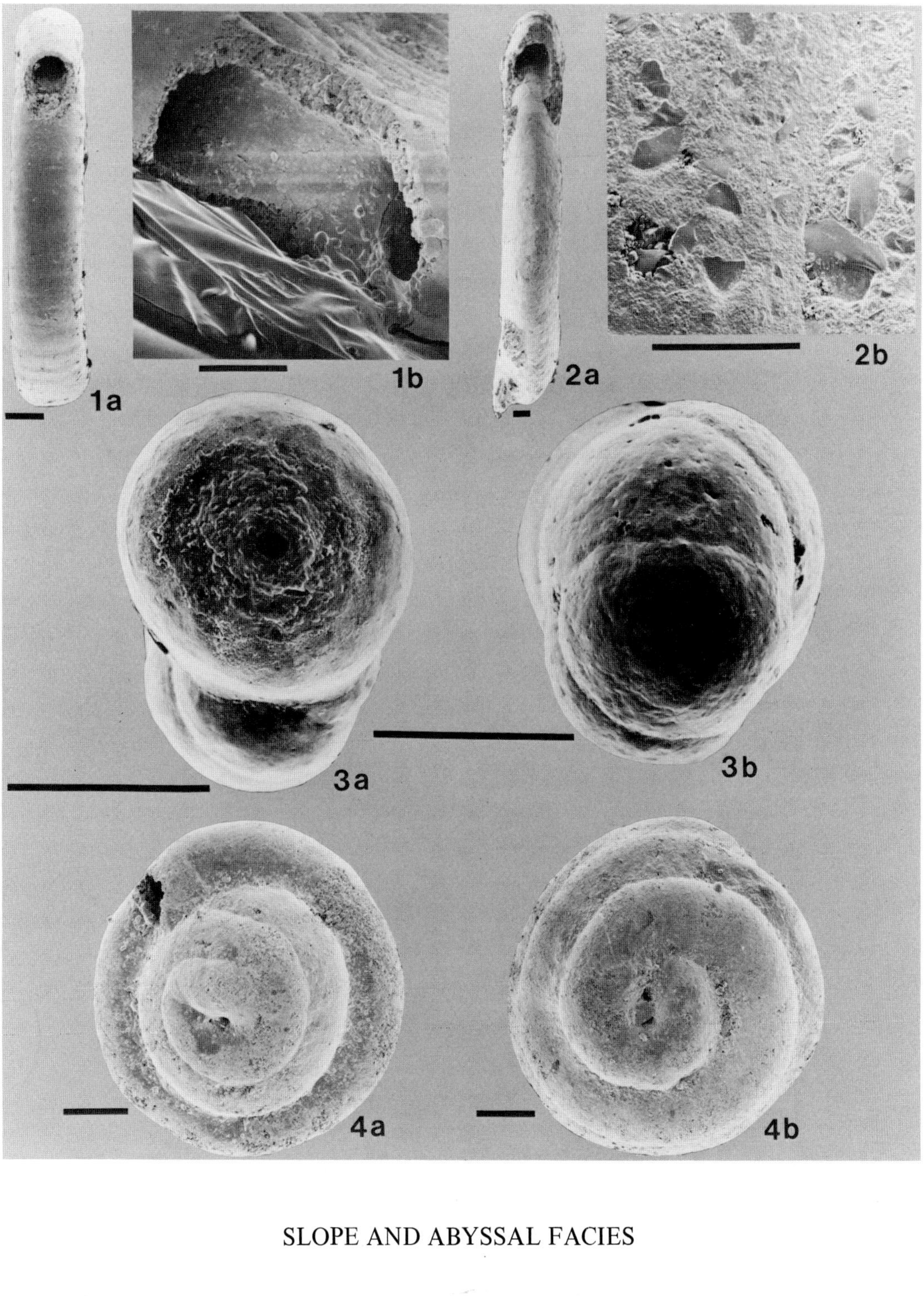

SLOPE AND ABYSSAL FACIES

PLATE 9

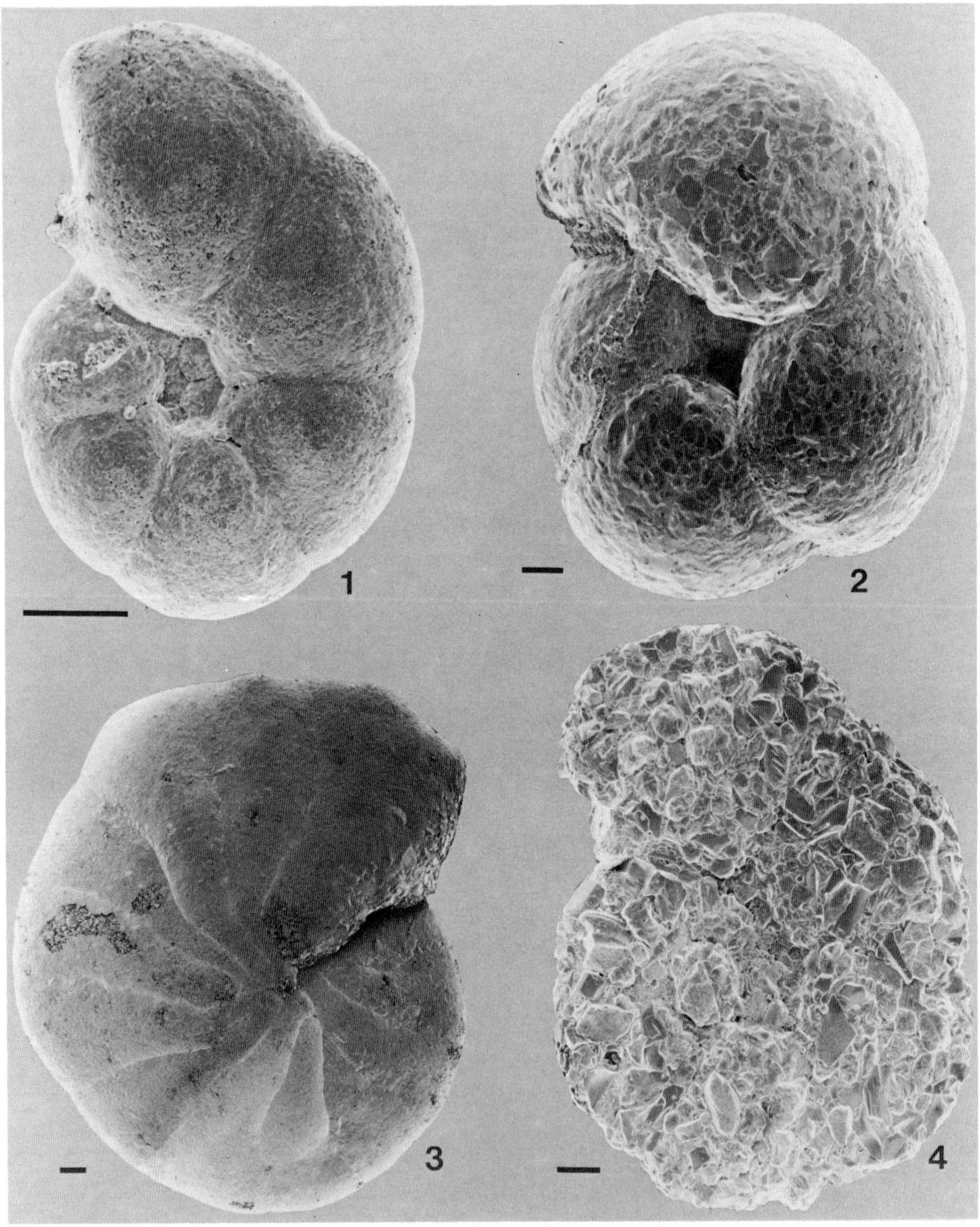

CRIBROSTOMOIDES, DISCAMMINA

PLATE 10

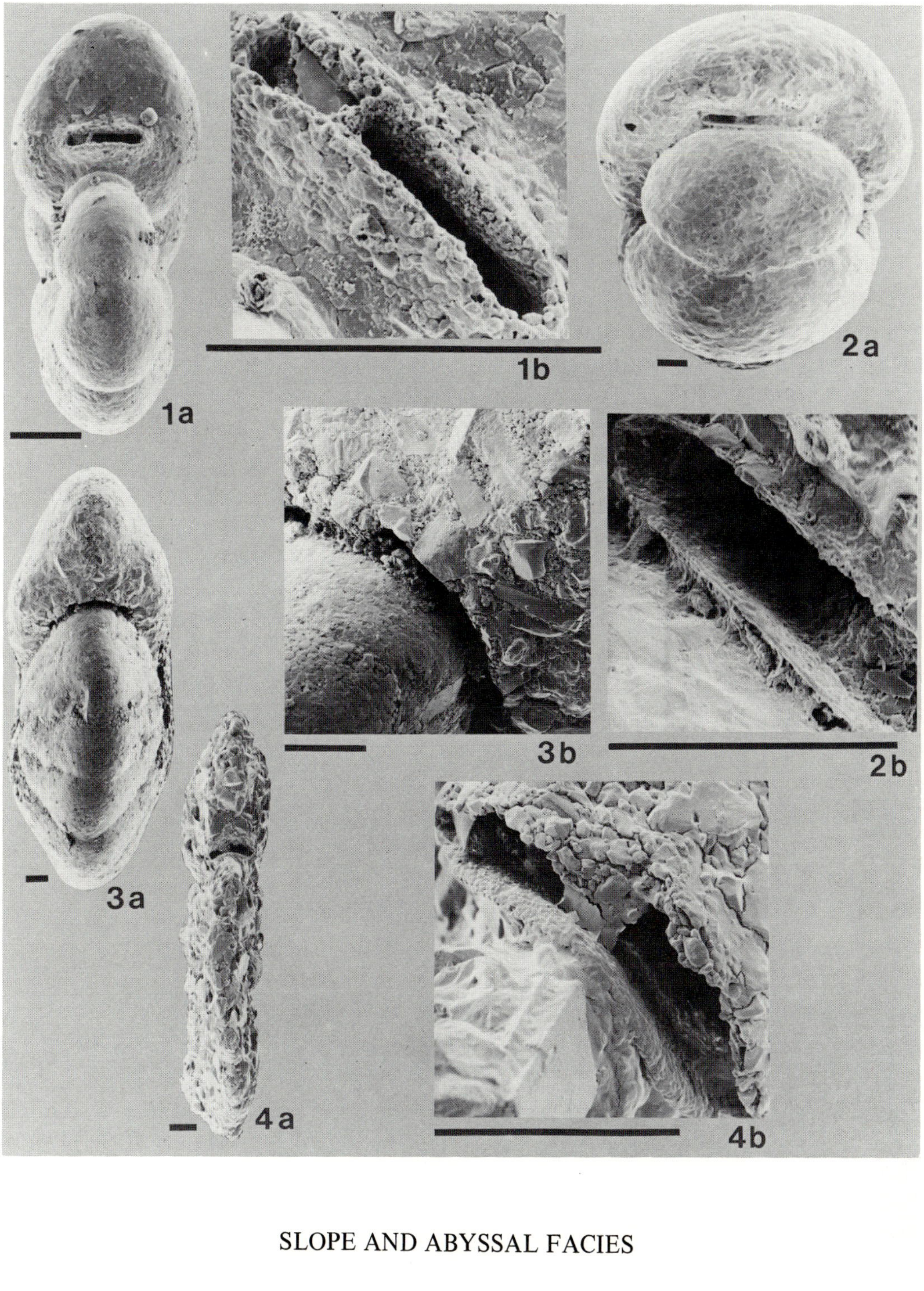

SLOPE AND ABYSSAL FACIES

PLATE 11

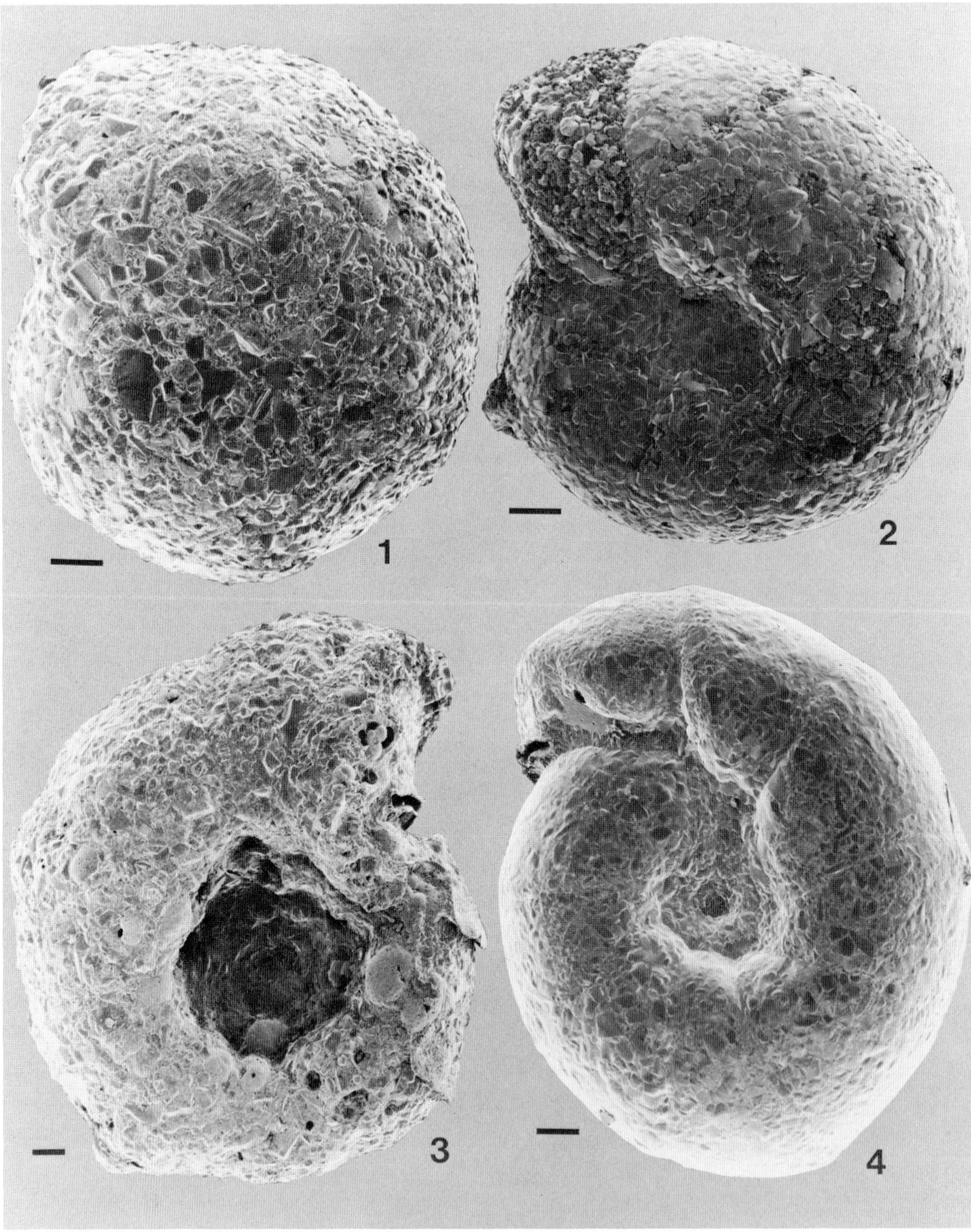

CRIBROSTOMOIDES

PLATE 12

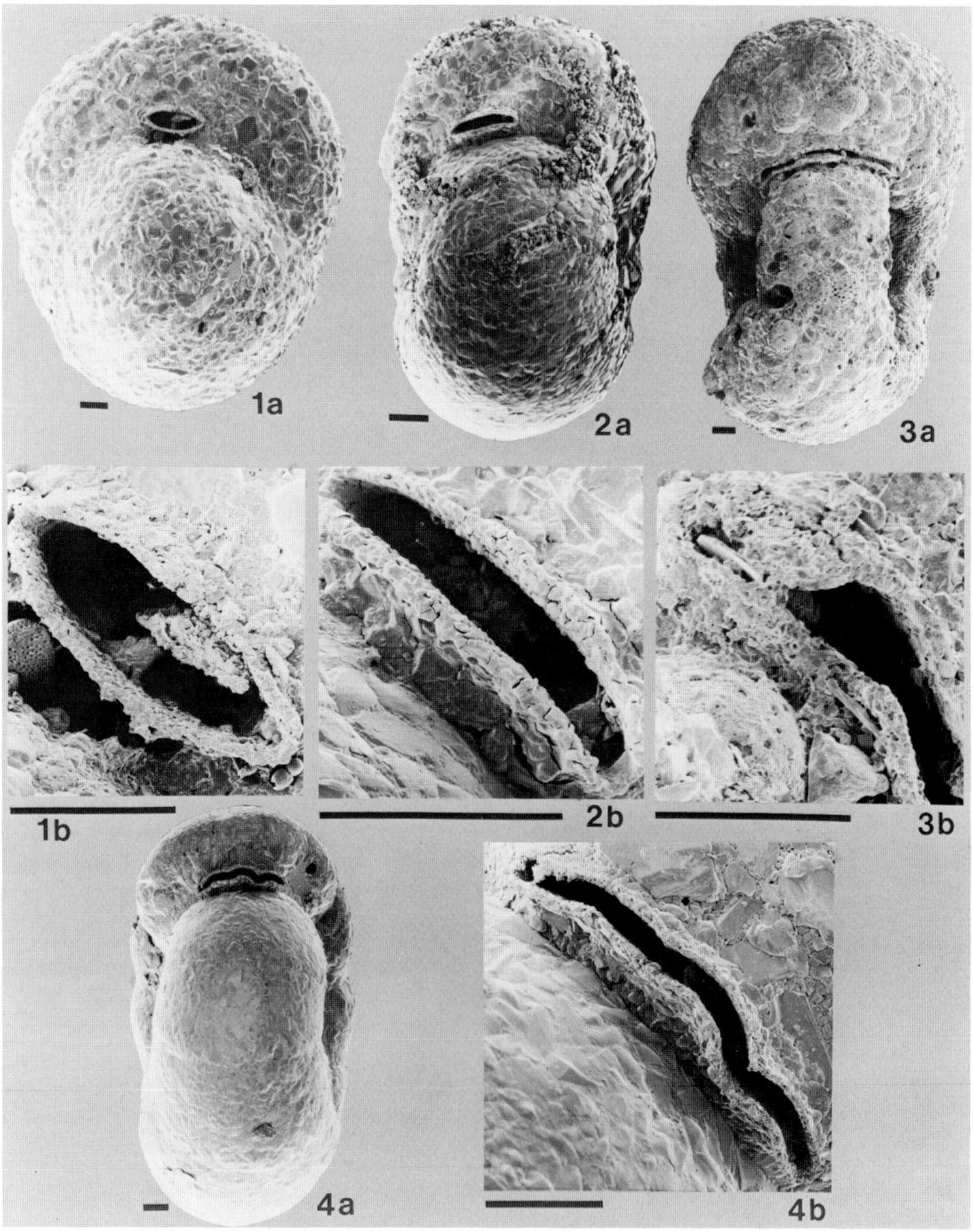

SLOPE AND ABYSSAL FACIES

PLATE 13

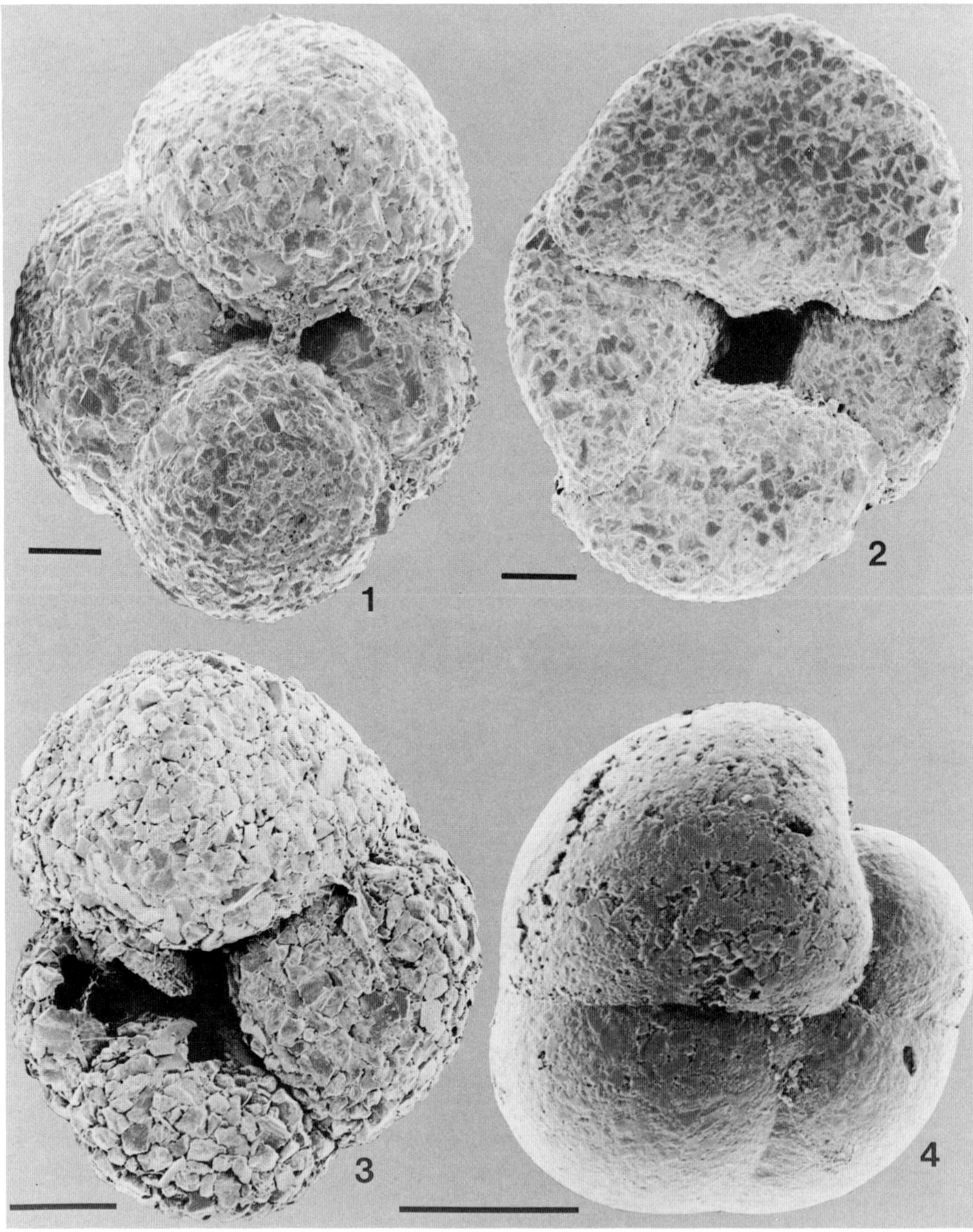

TROCHAMMINA, SIPHOTROCHAMMINA, HAPLOPHRAGMOIDES

PLATE 14

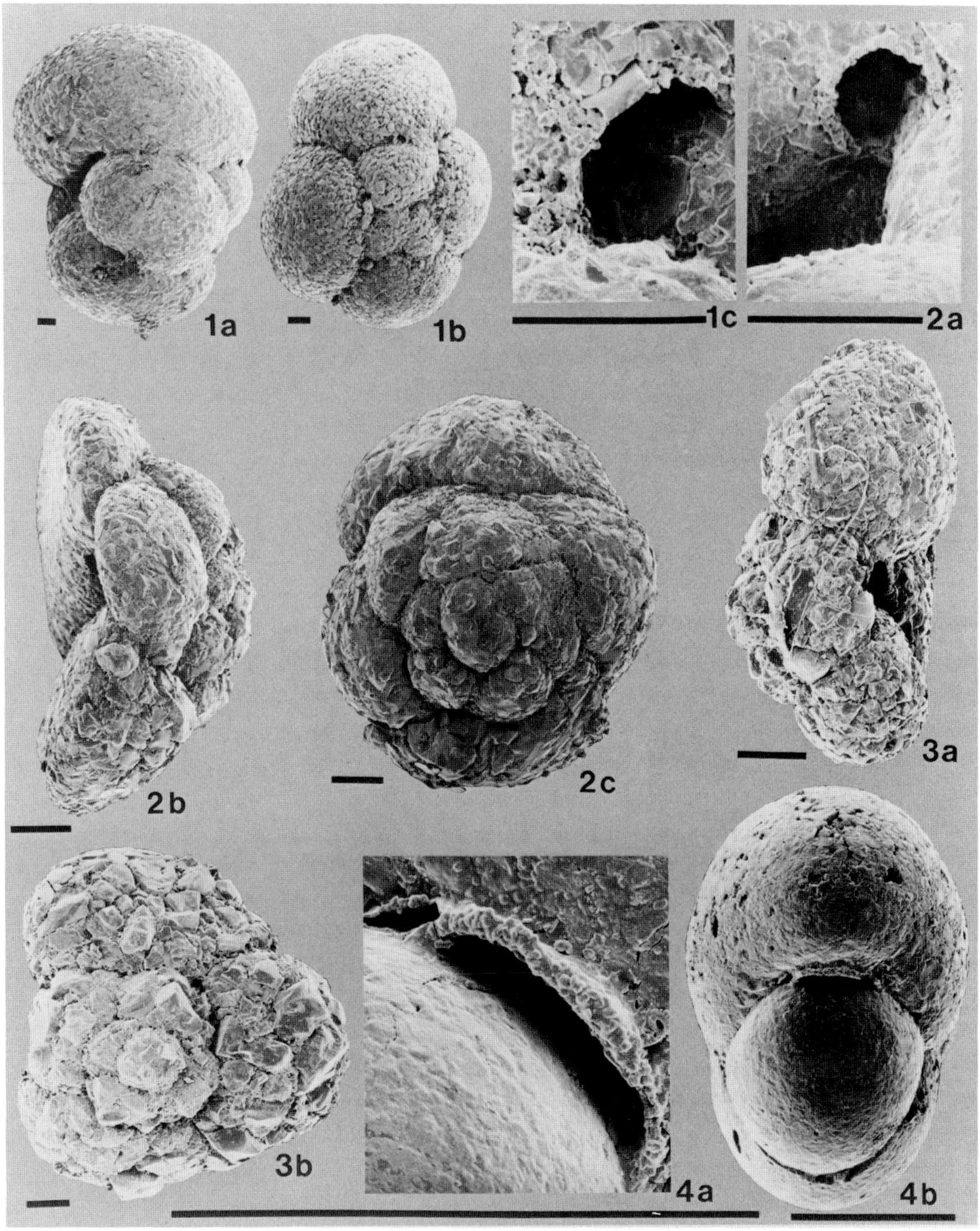

SLOPE AND ABYSSAL FACIES

PLATE 15

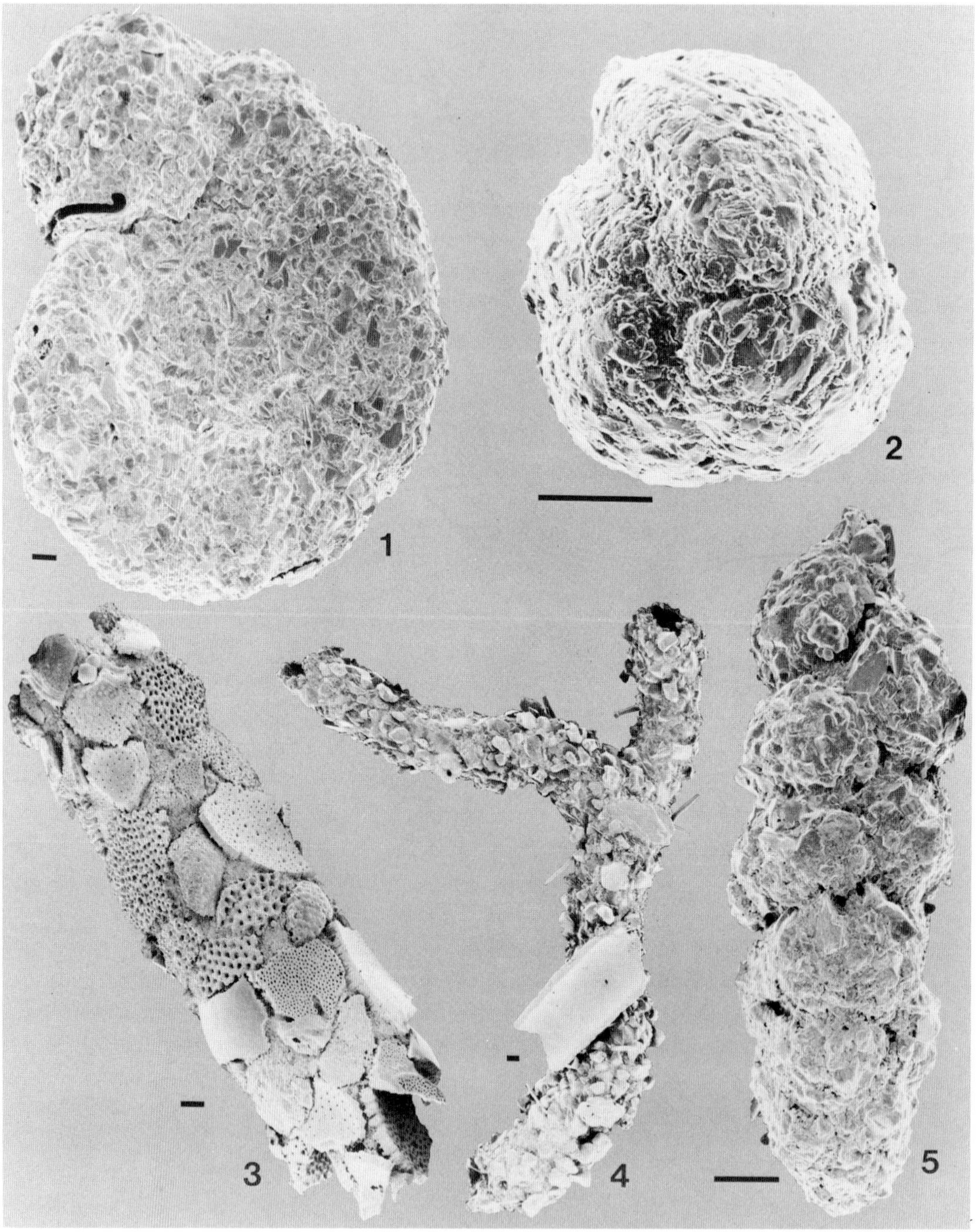

ALVEOLOPHRAGMIUM, ADERCOTRYMA, BATHYSIPHON, SACCORHIZA, KARRERIELLA

PLATE 16

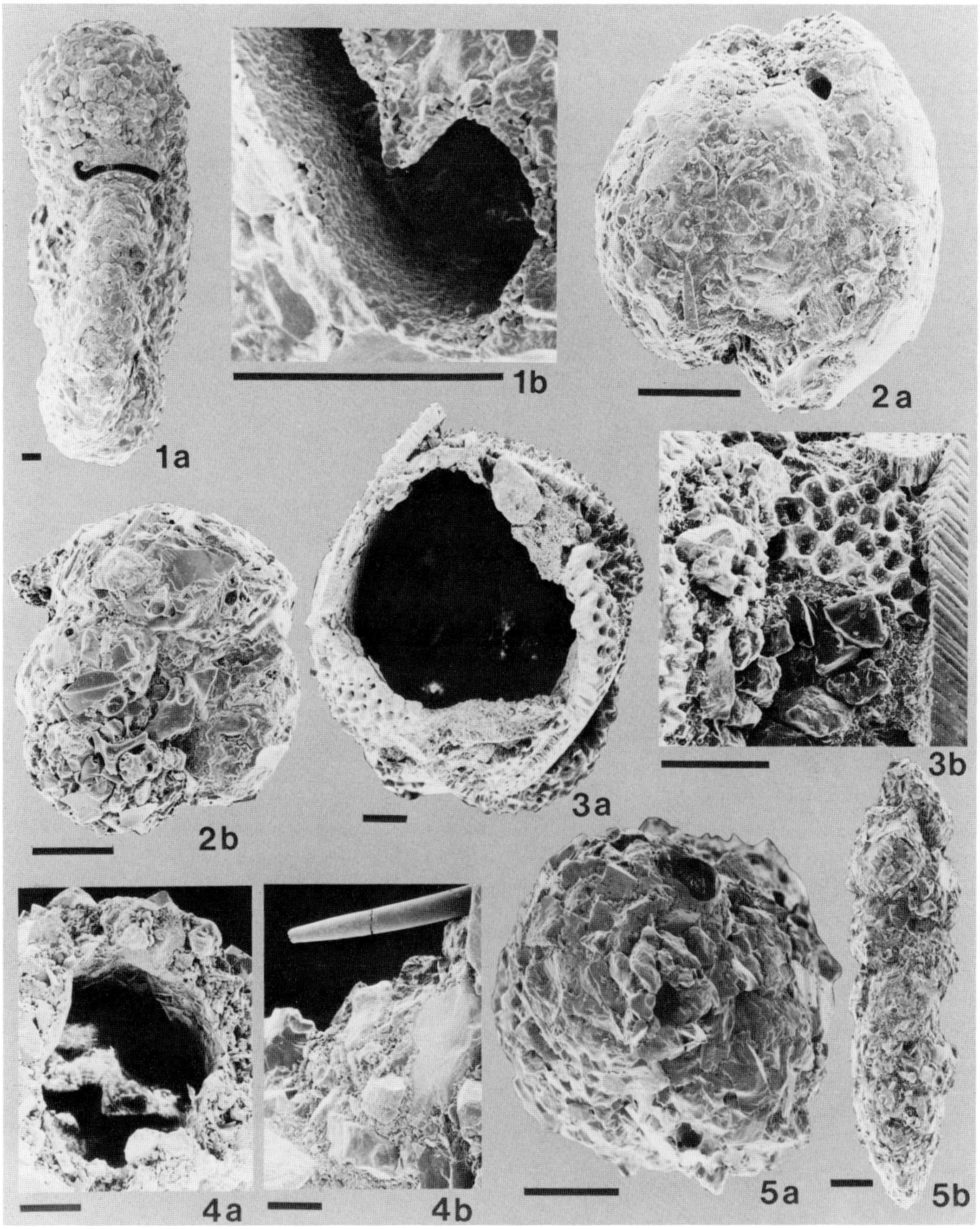

SLOPE AND ABYSSAL FACIES

PLATE 17

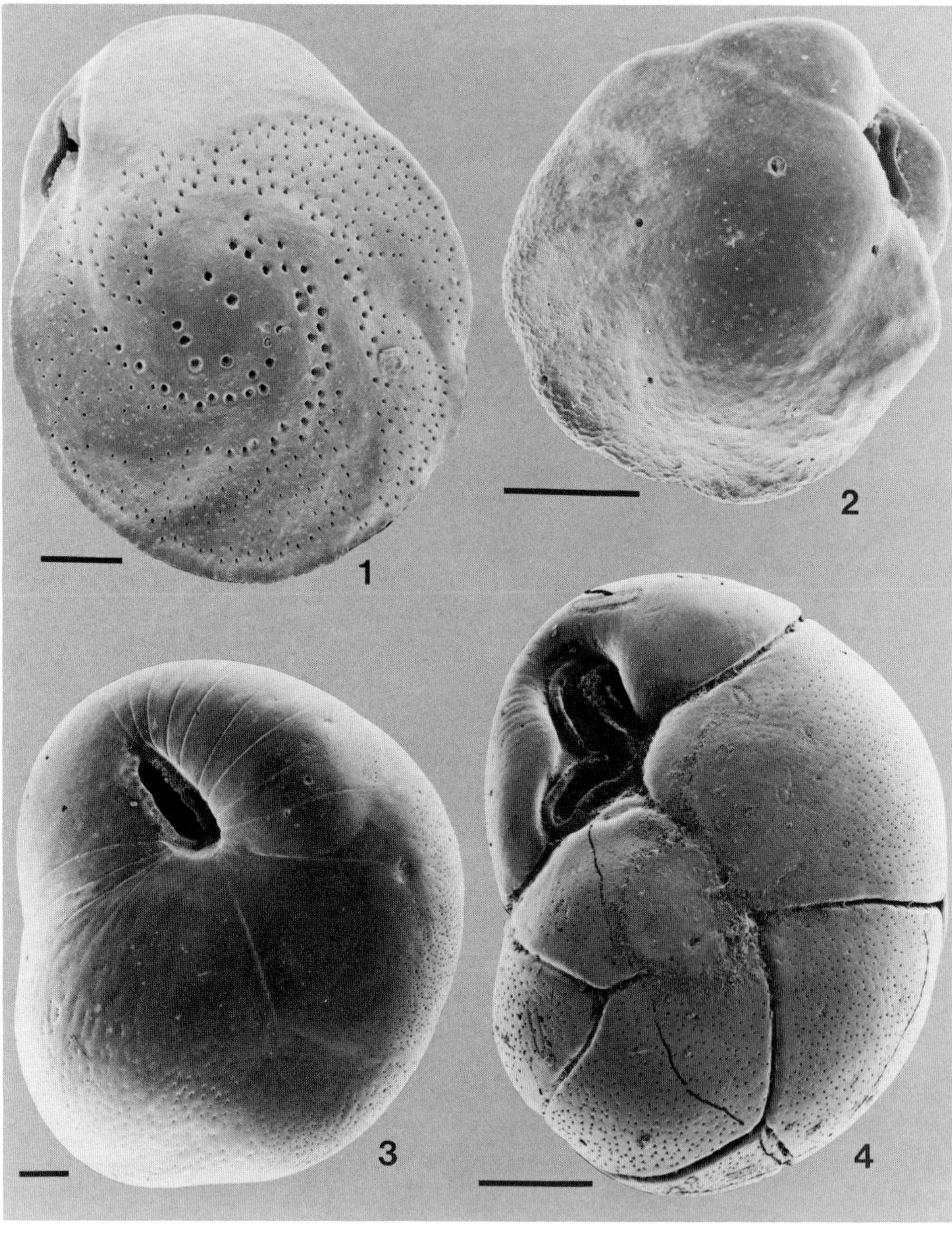

ISLANDIELLA

PLATE 18

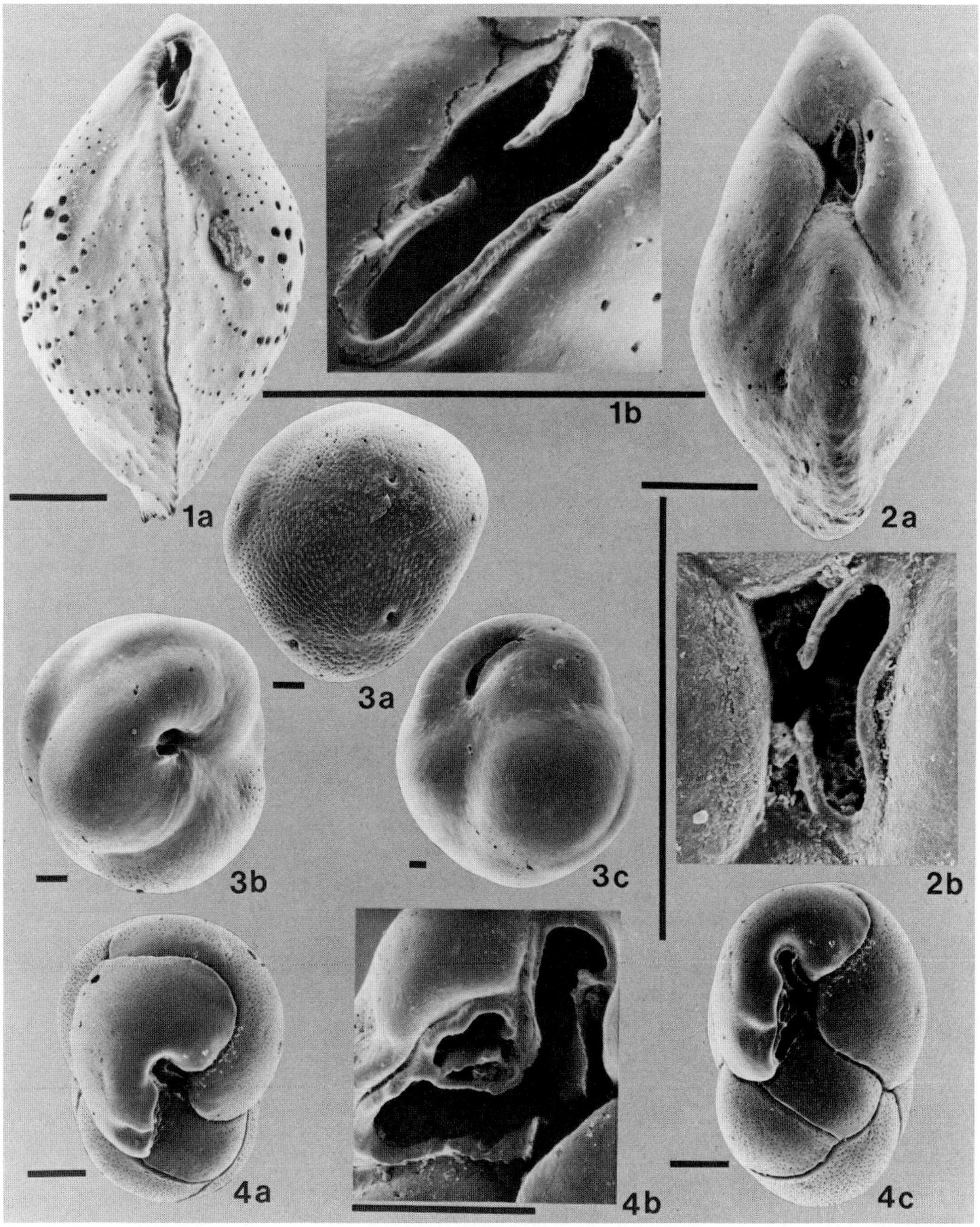

SHELF, SLOPE, AND ABYSSAL FACIES

PLATE 19

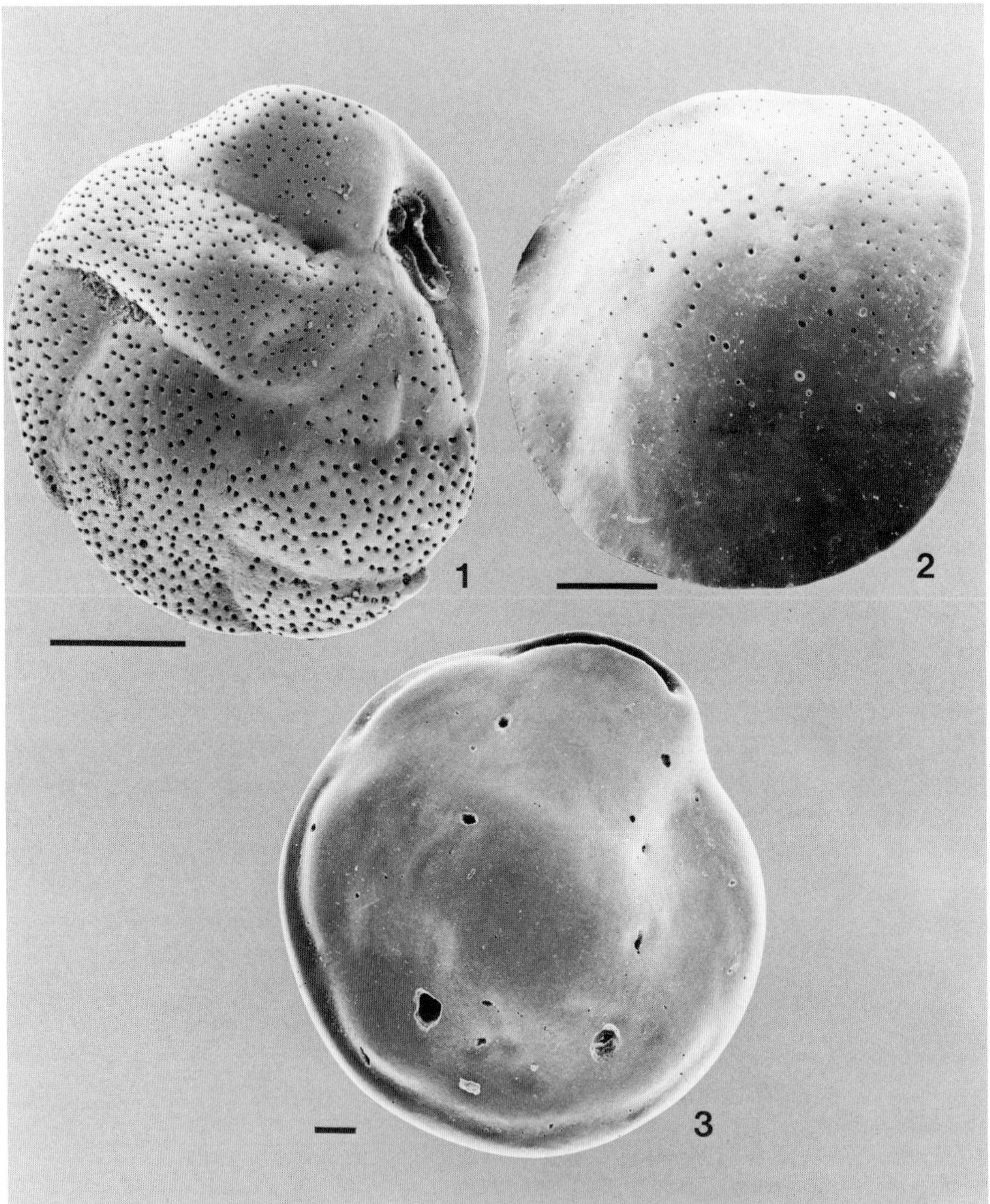

ISLANDIELLA, HOEGLUNDINA

PLATE 20

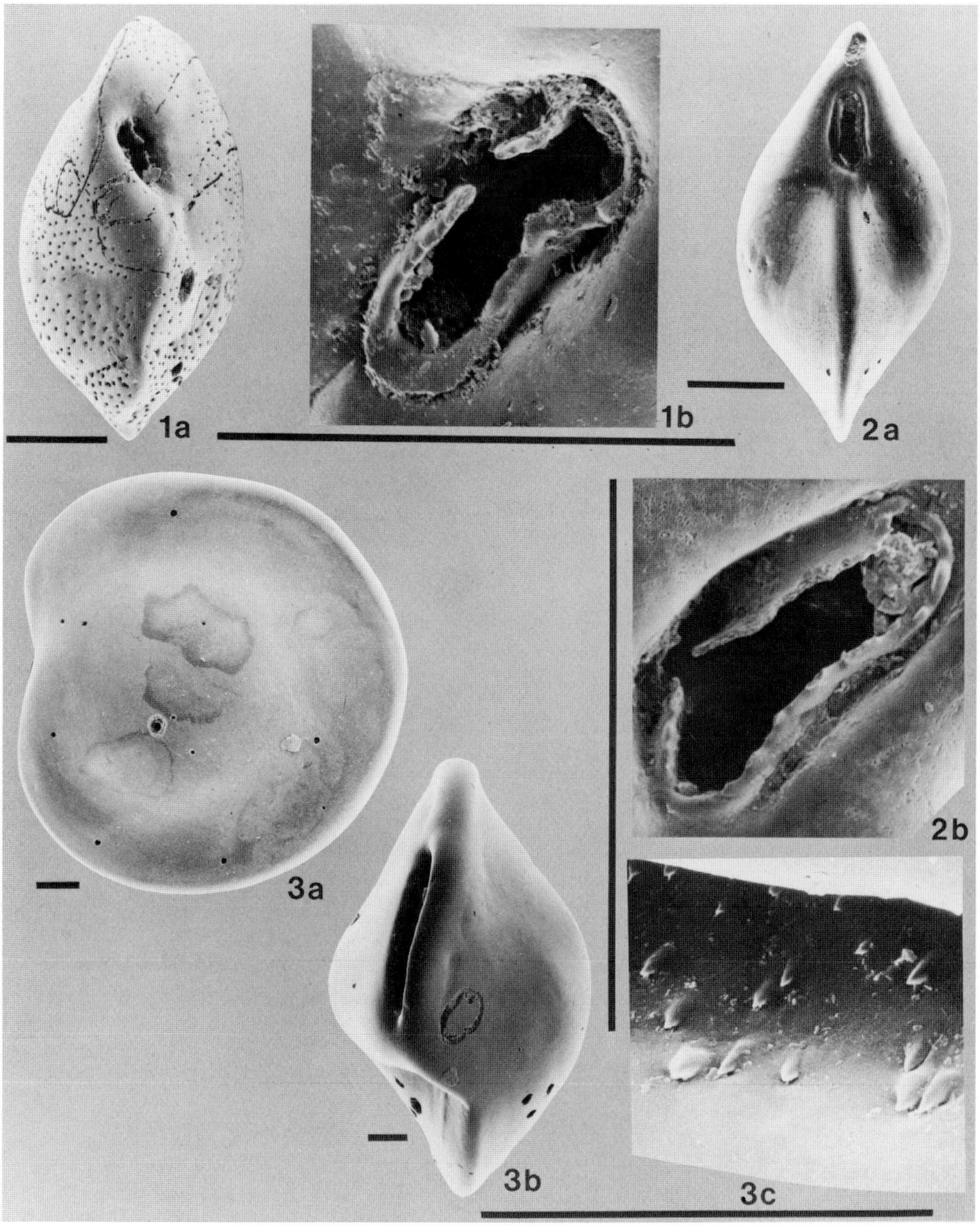

SHELF, SLOPE, AND ABYSSAL FACIES

PLATE 21

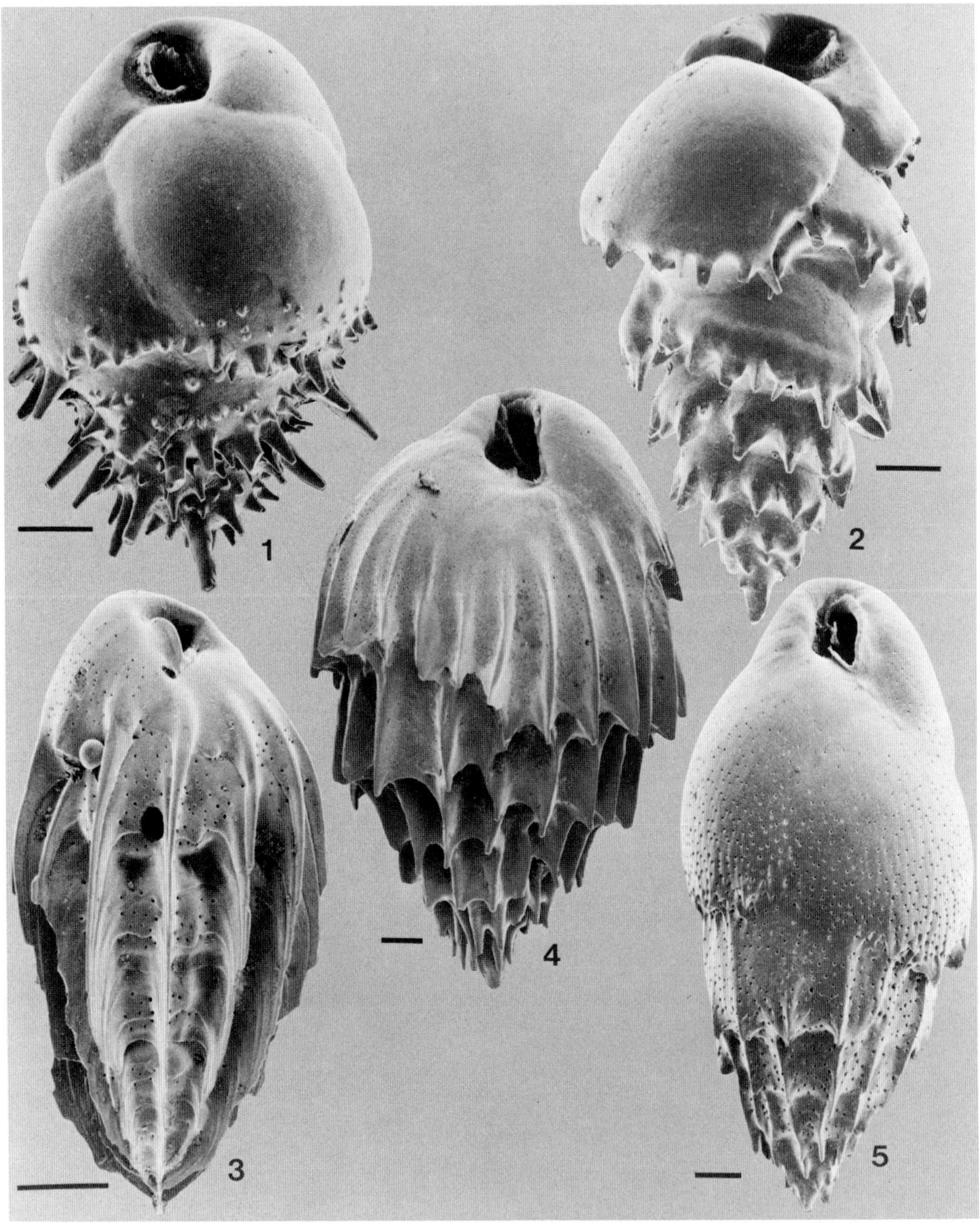

BULIMINA

PLATE 22

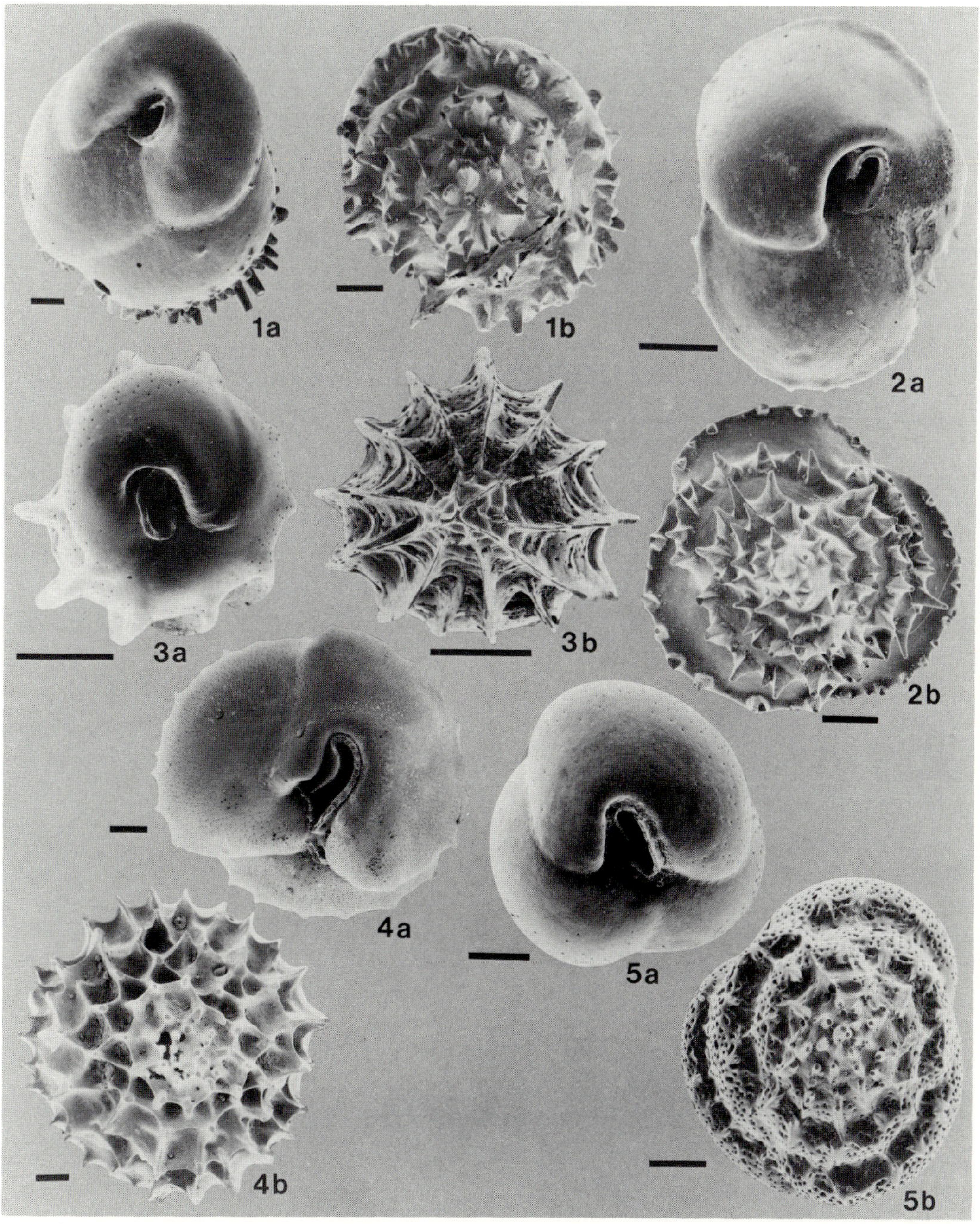

SHELF AND SLOPE FACIES

PLATE 23

BRIZALINA

PLATE 24

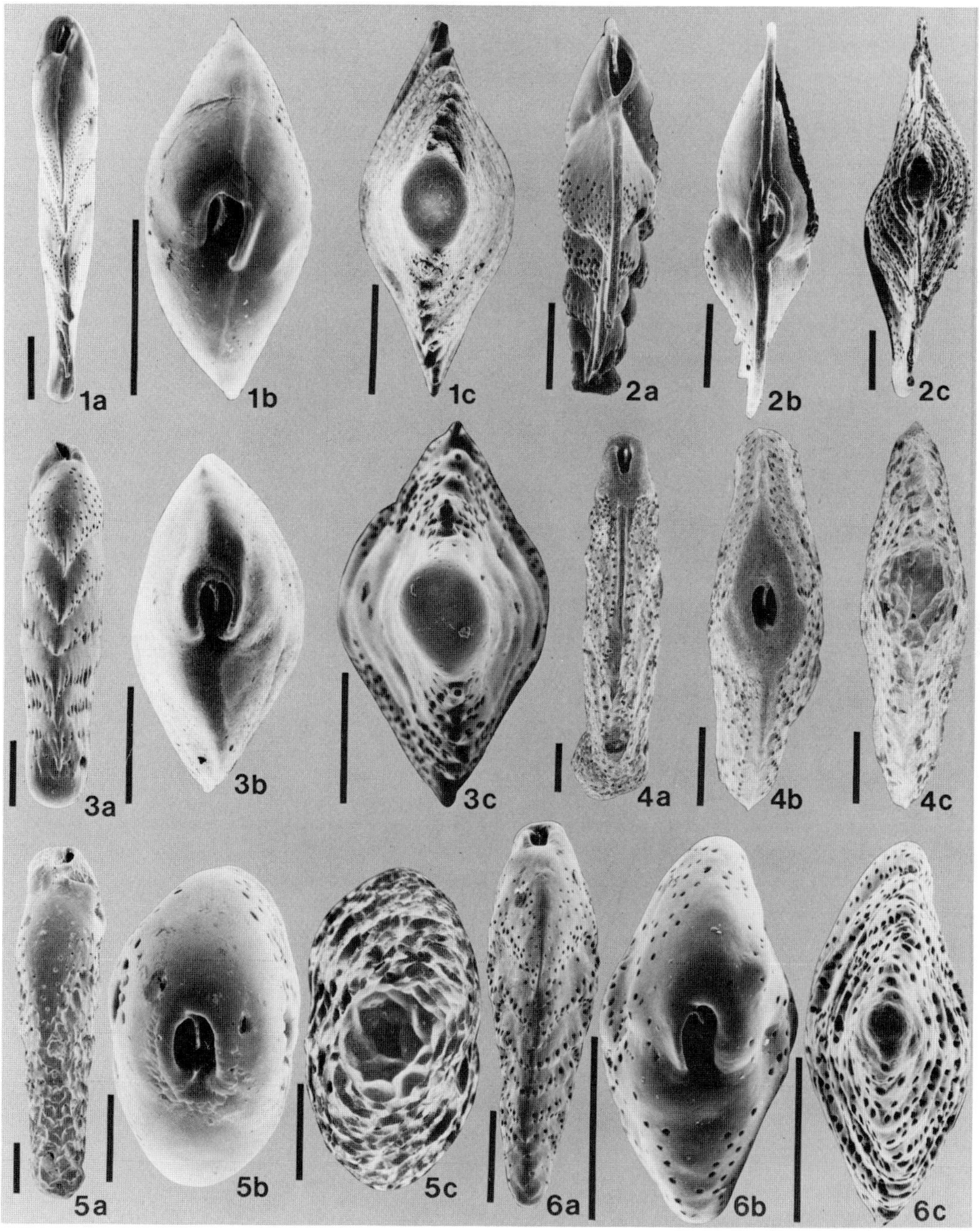

SHELF AND SLOPE FACIES

PLATE 25

BRIZALINA

PLATE 26

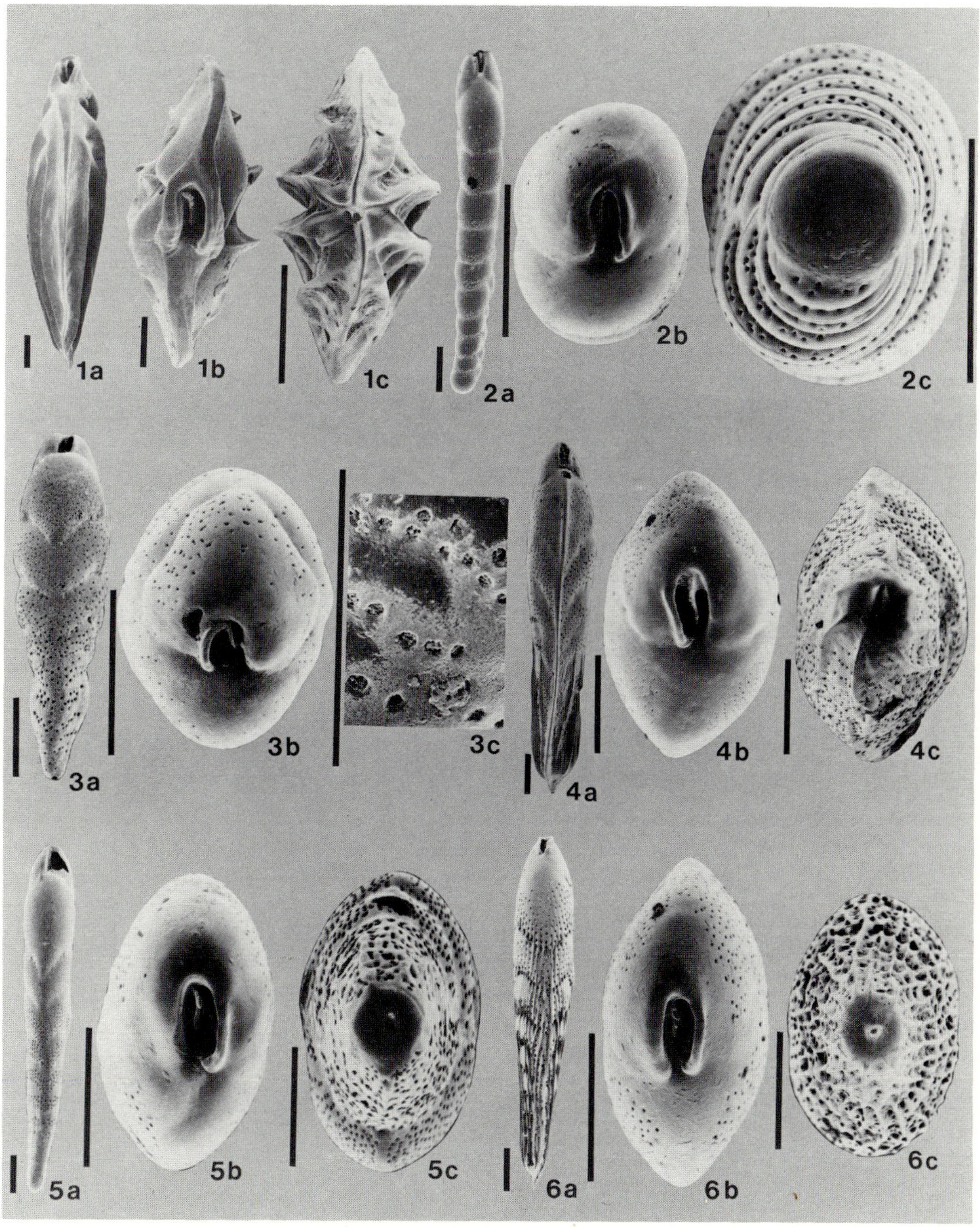

SHELF AND SLOPE FACIES

PLATE 27

UVIGERINA

PLATE 28

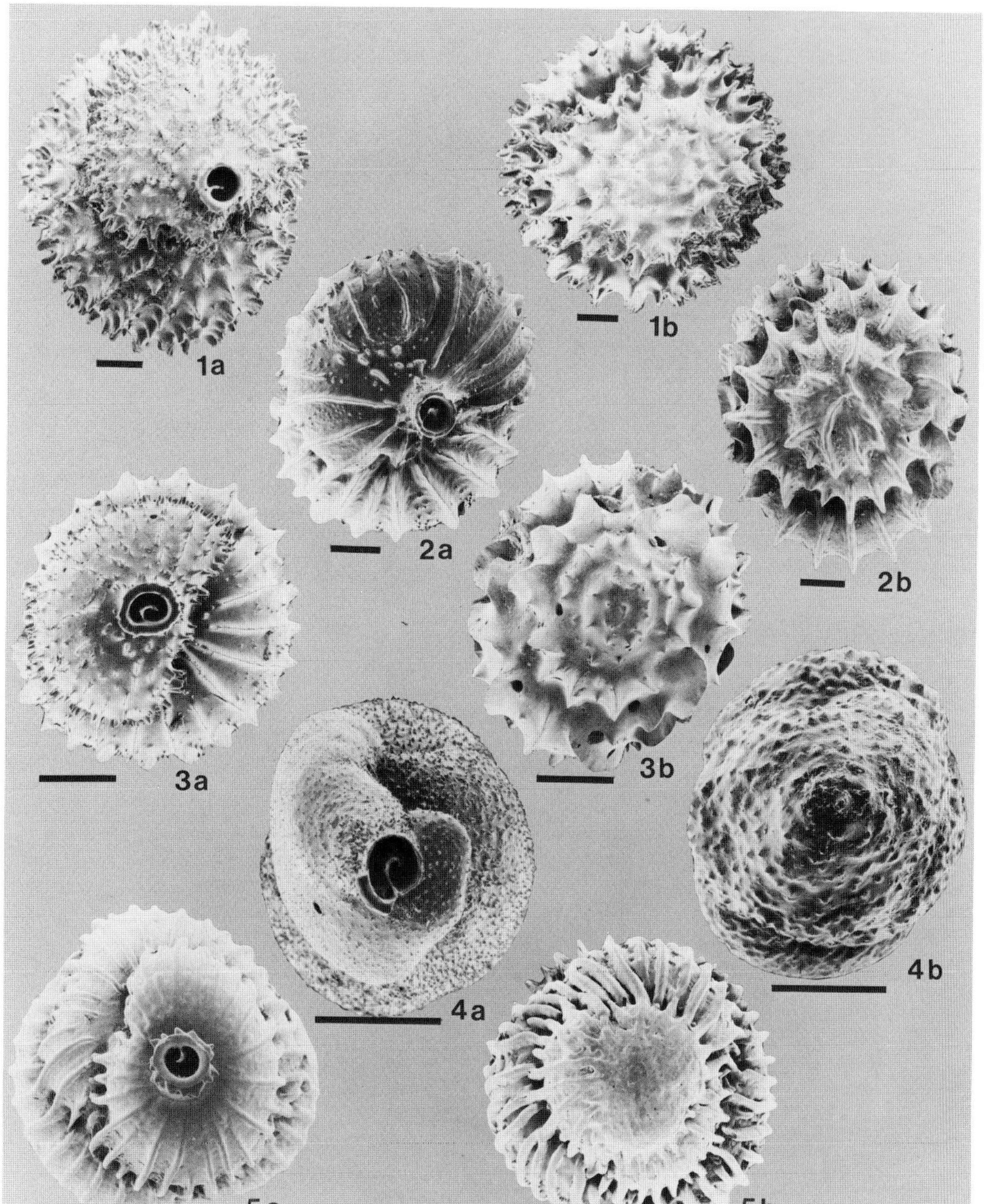

SHELF AND SLOPE FACIES

PLATE 29

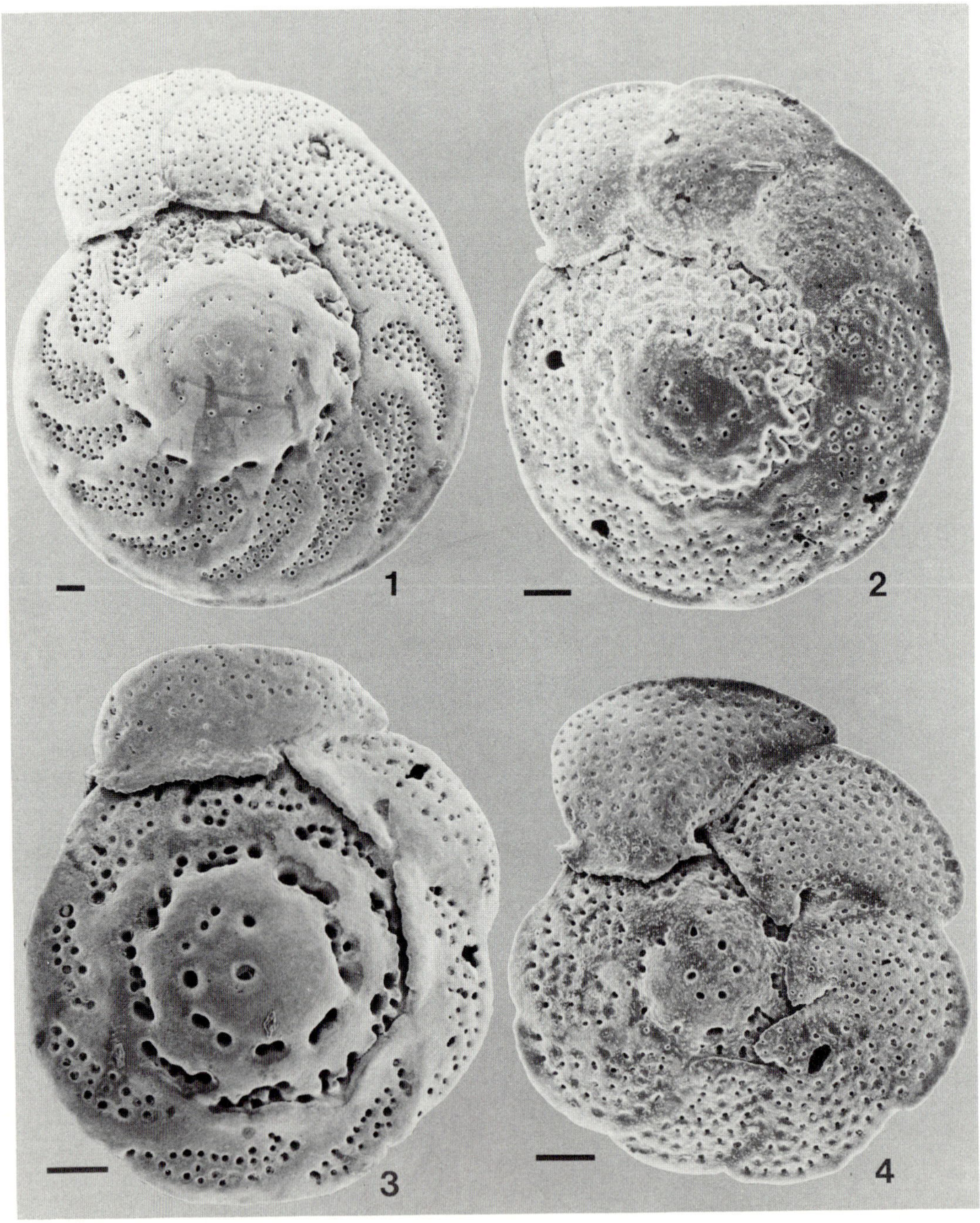

1. "floridanus" f. bathyalis = pachyderma f. bathyalis
2. "floridanus" f. sublittoralis = pachyderma f. sublittoralis

CIBICIDOIDES

3. sp. A

4. mollis

PLATE 30

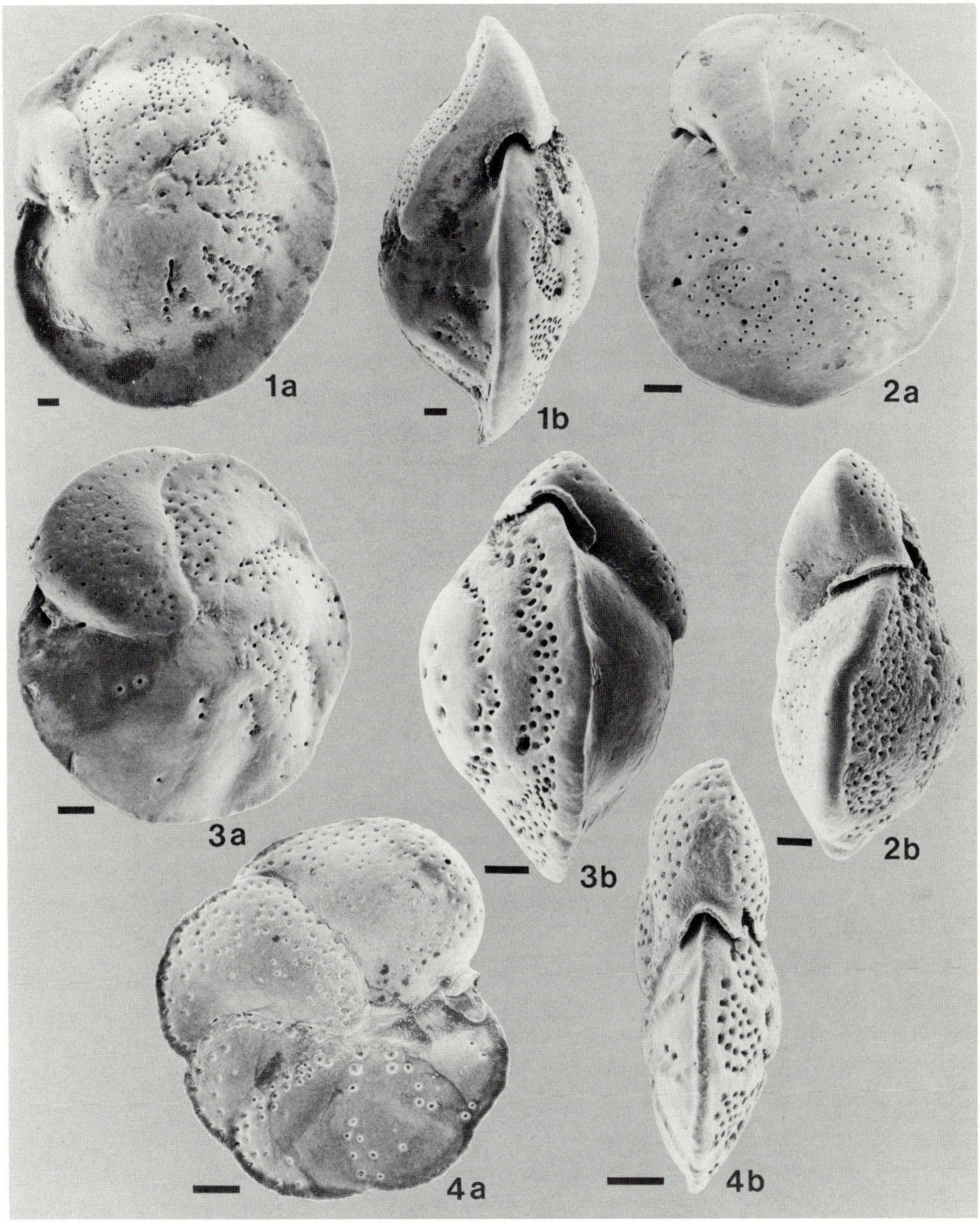

SHELF FACIES

PLATE 31

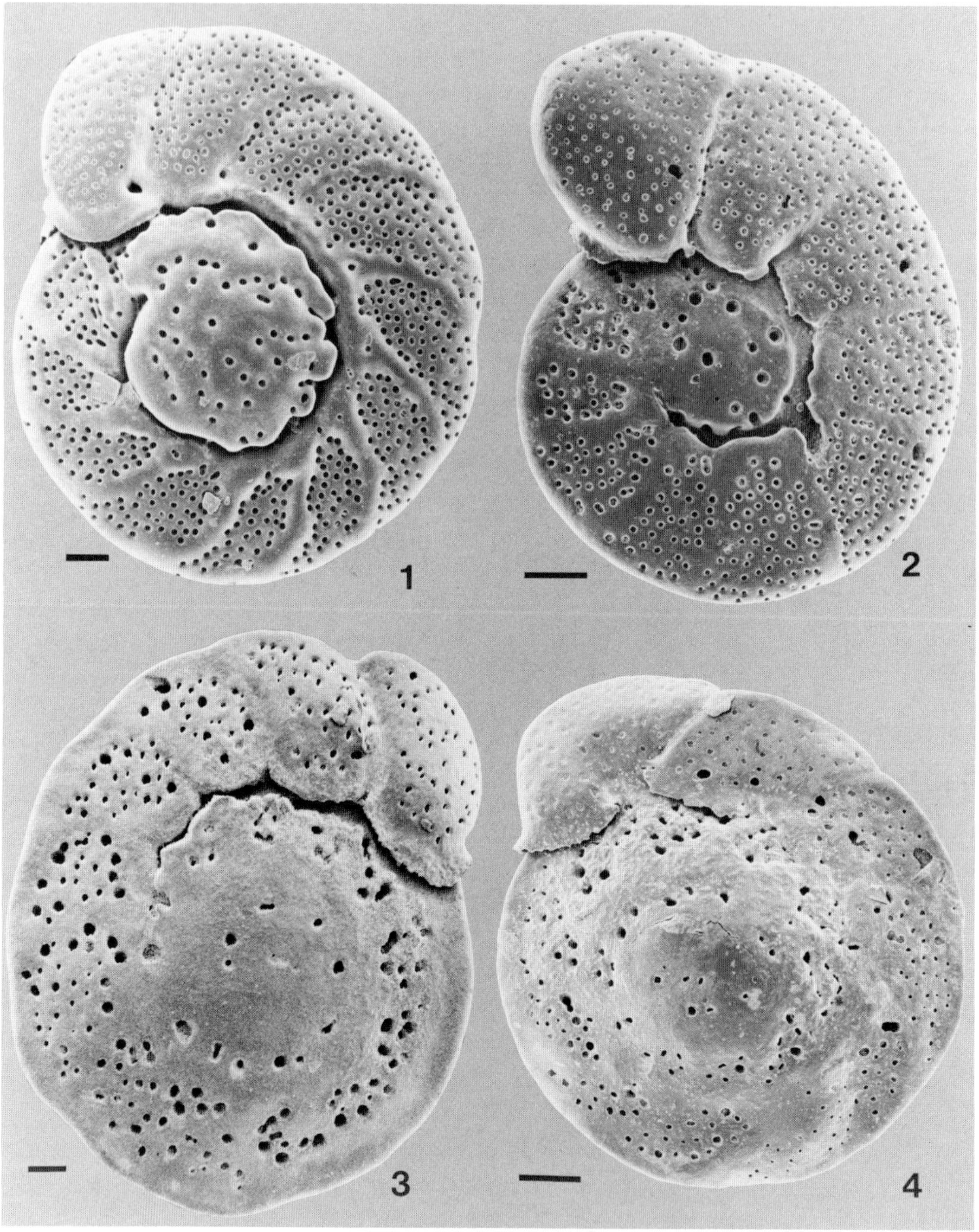

CIBICIDOIDES

PLATE 32

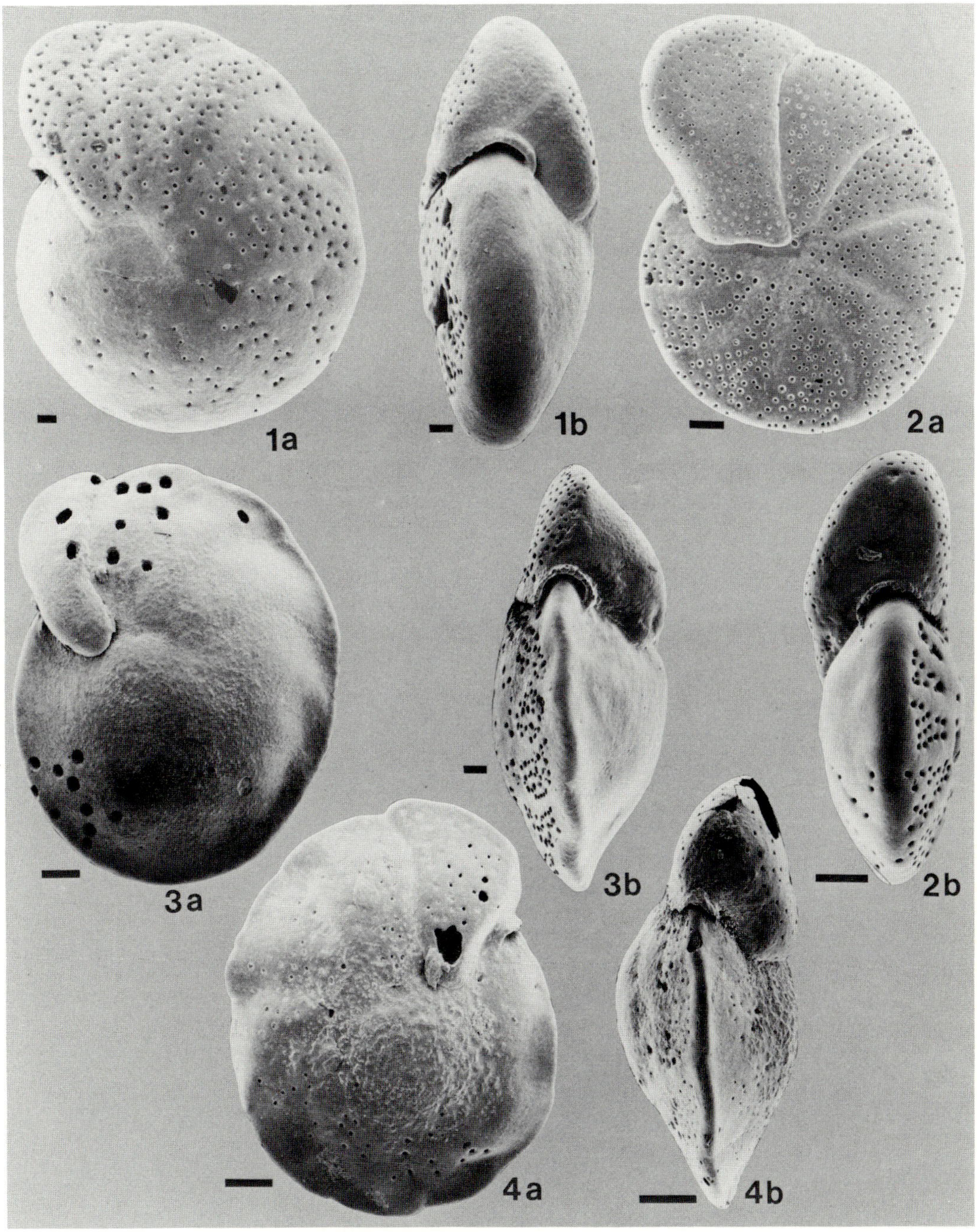

SLOPE FACIES

PLATE 33

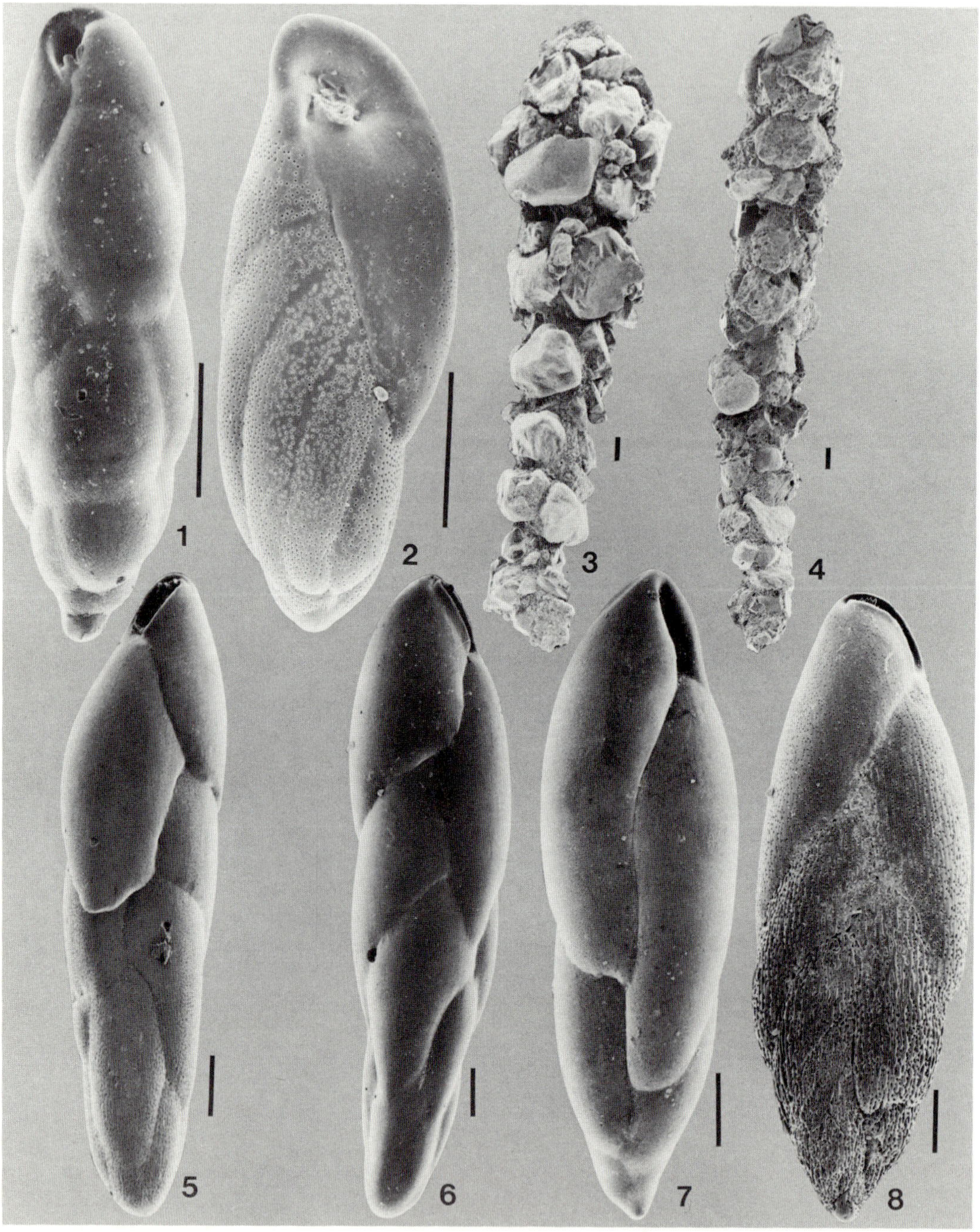

BULIMINELLA, BIGENERINA, FURSENKOINA

PLATE 34

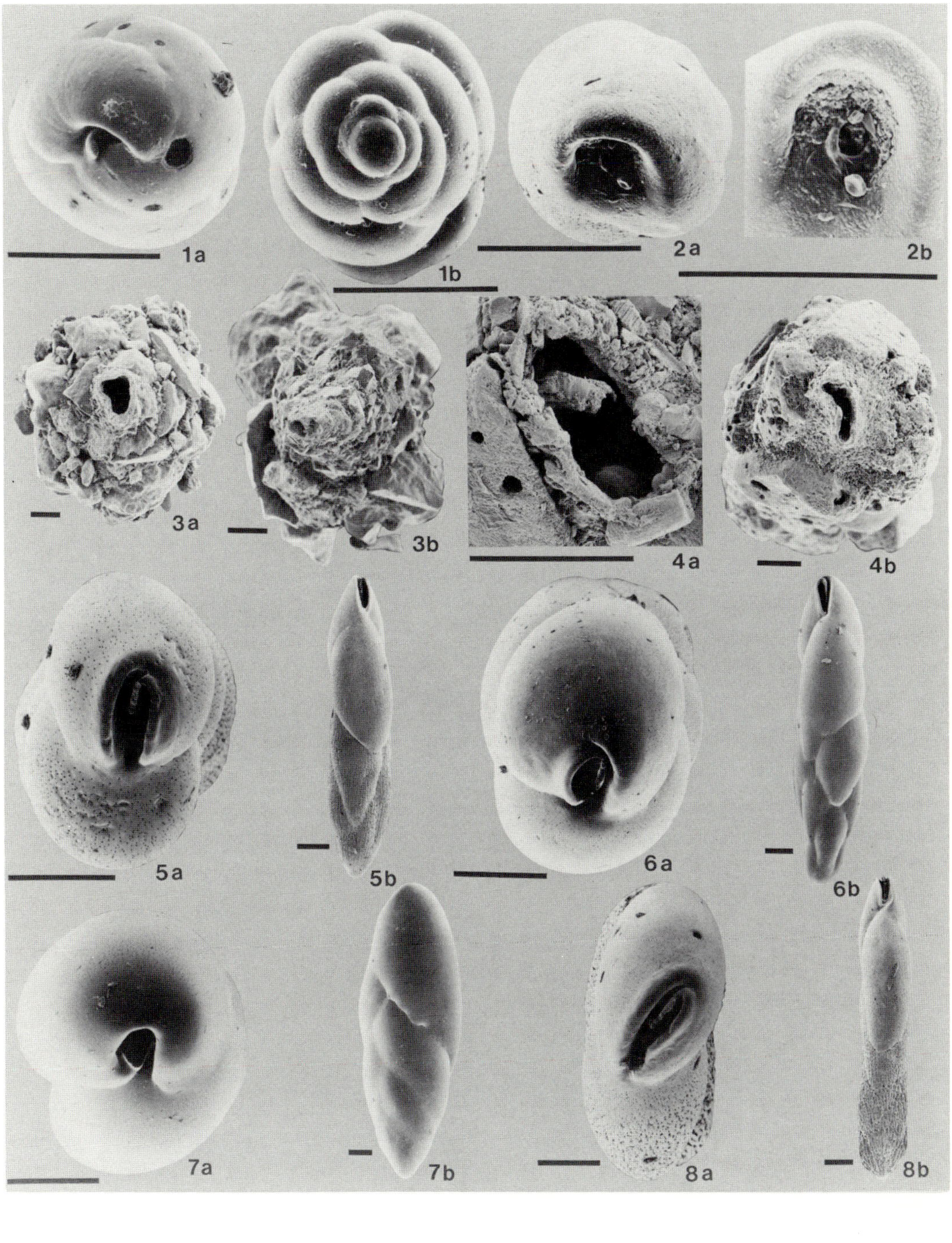

SHELF FACIES

PLATE 35

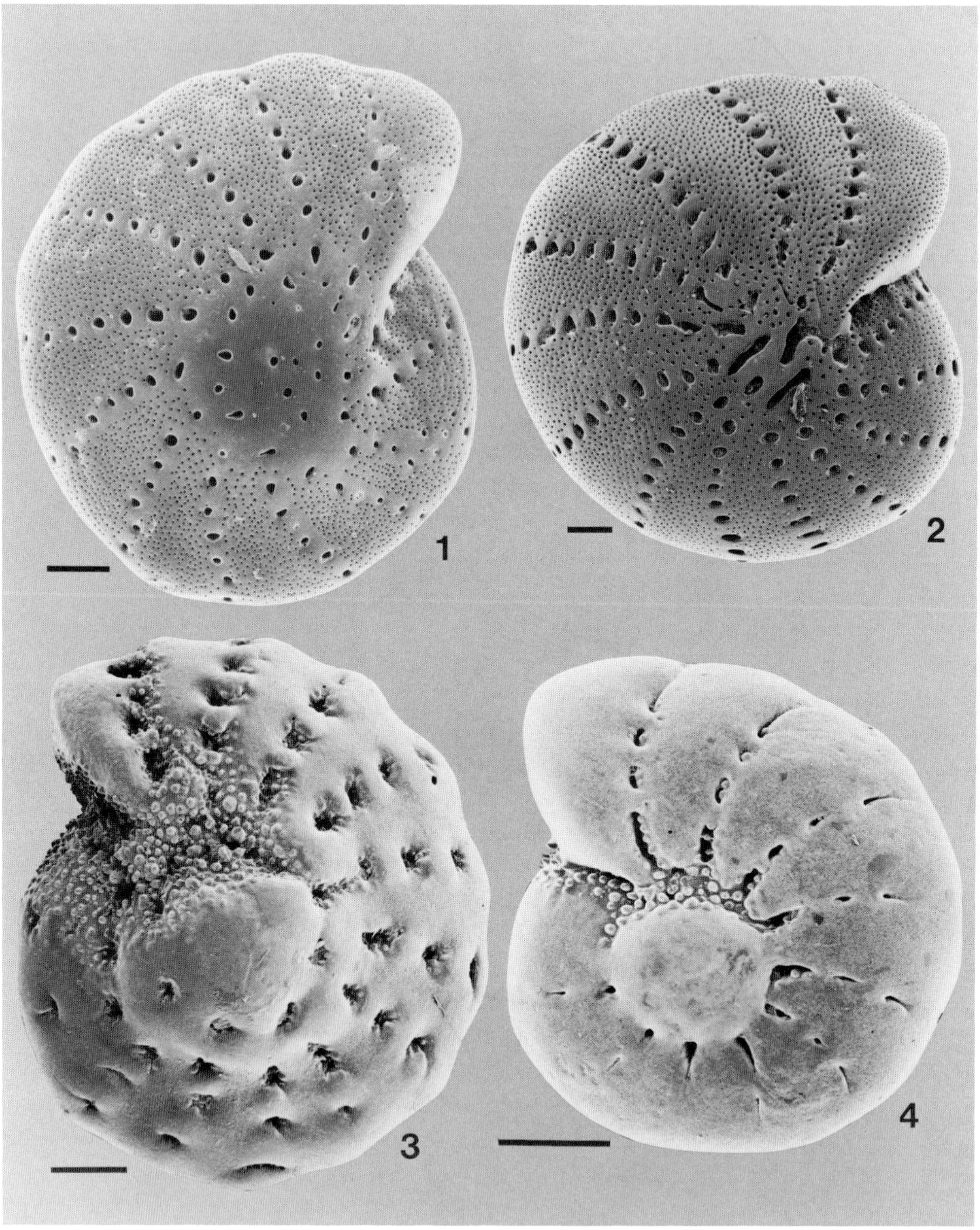

ELPHIDIUM

PLATE 36

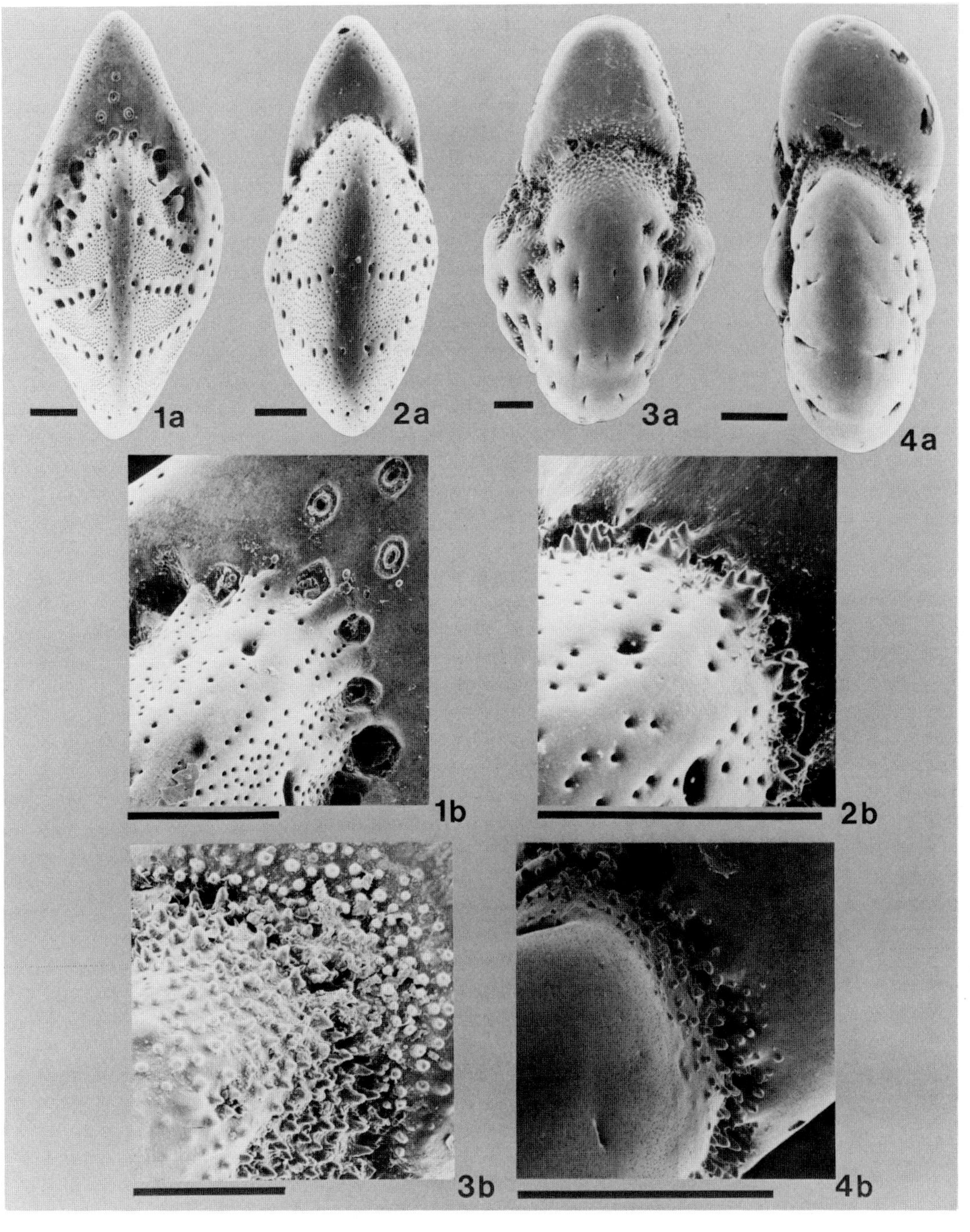

ESTUARY AND SHELF FACIES

PLATE 37

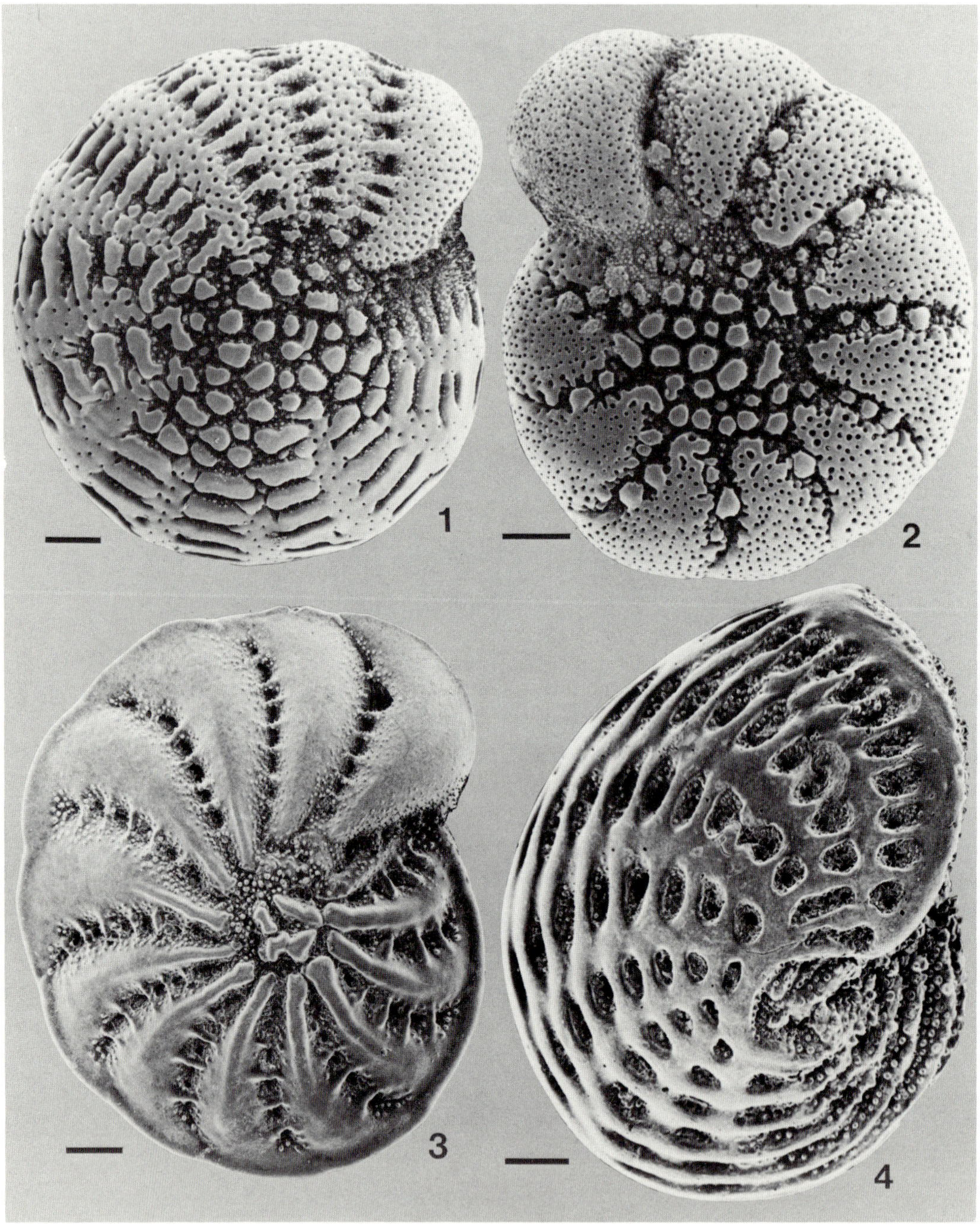

ELPHIDIUM

PLATE 38

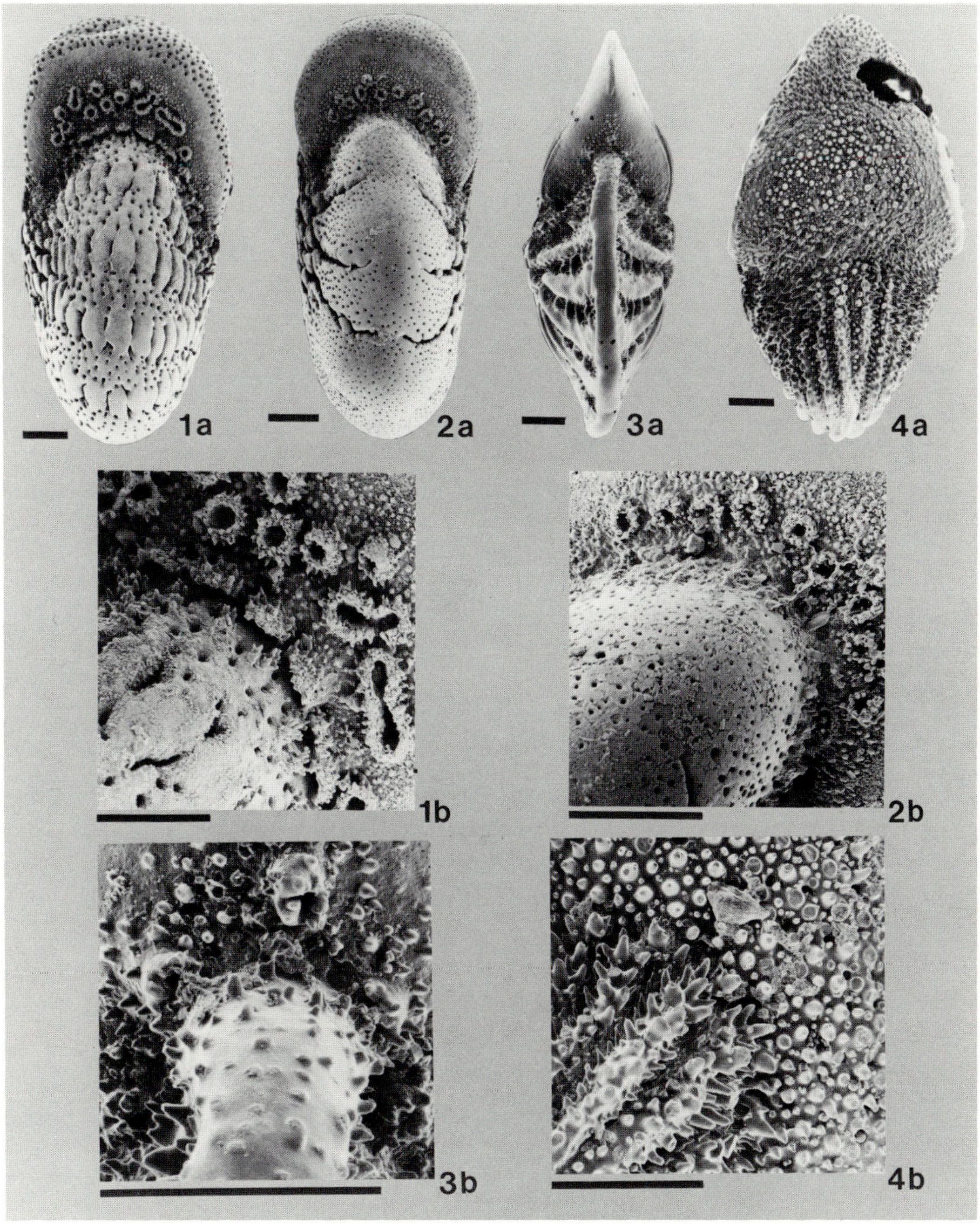

ESTUARY AND SHELF FACIES

PLATE 39

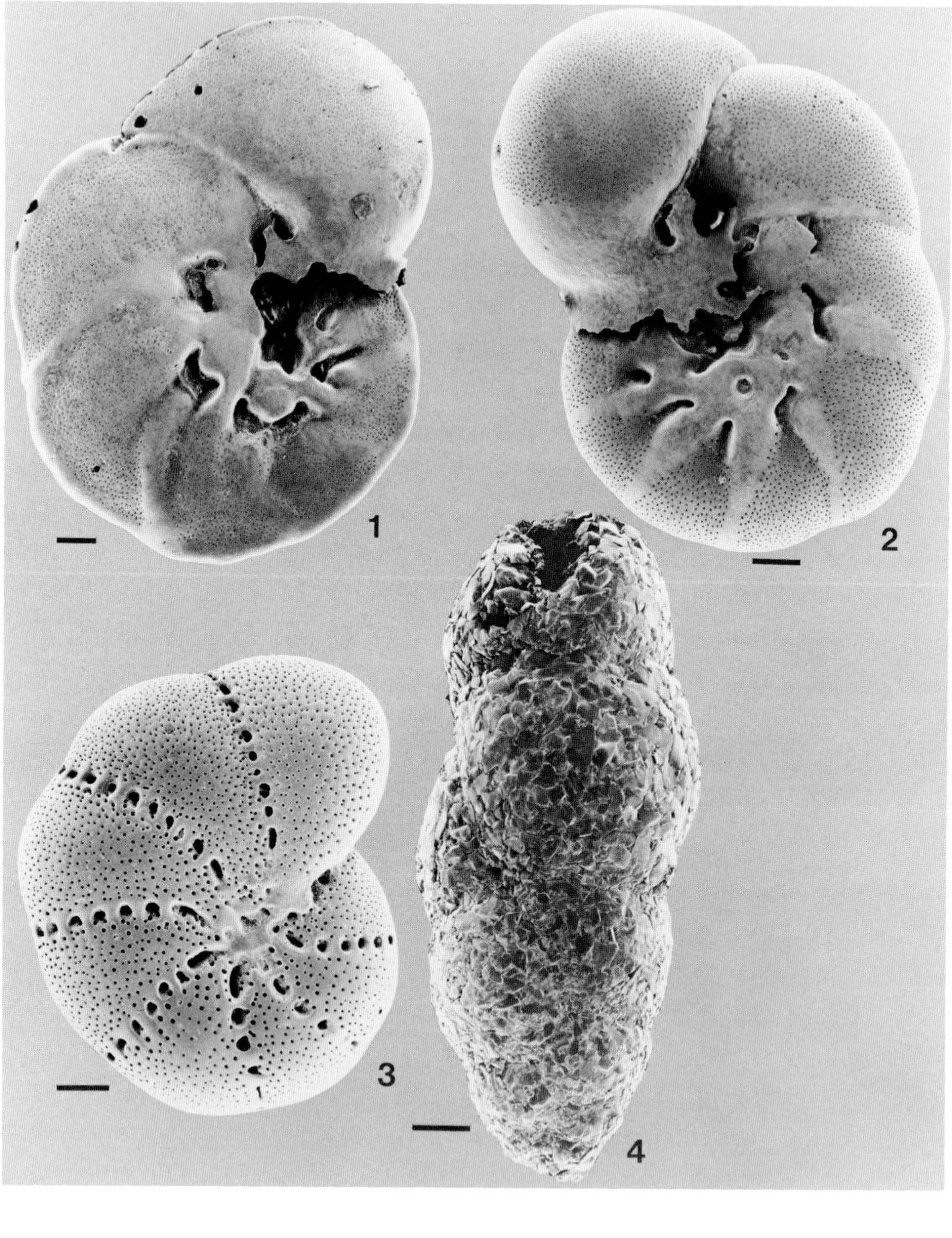

HANZAWAIA, ELPHIDIUM, "GOESELLA"

PLATE 40

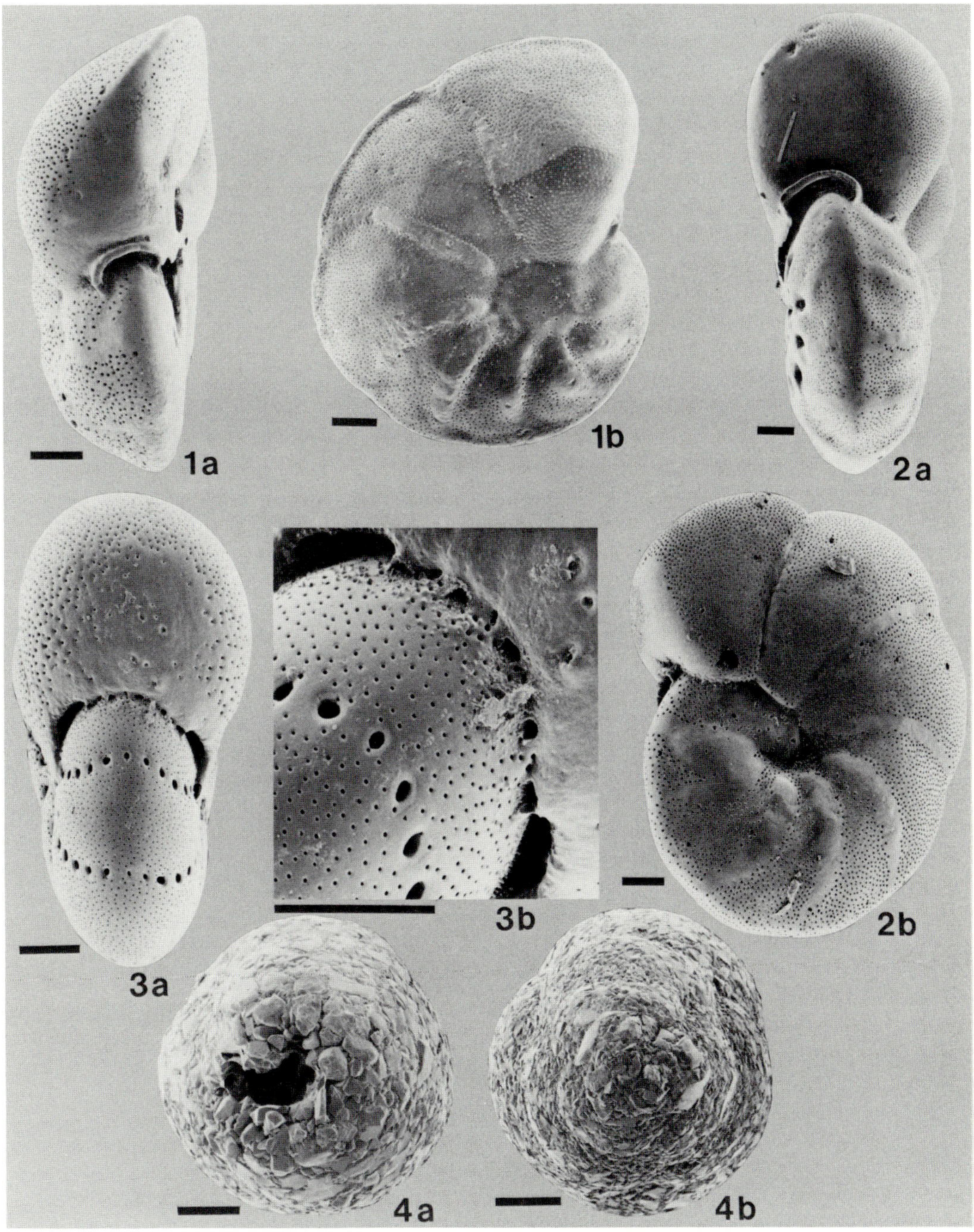

ESTUARY, SHELF, AND SLOPE FACIES

PLATE 41

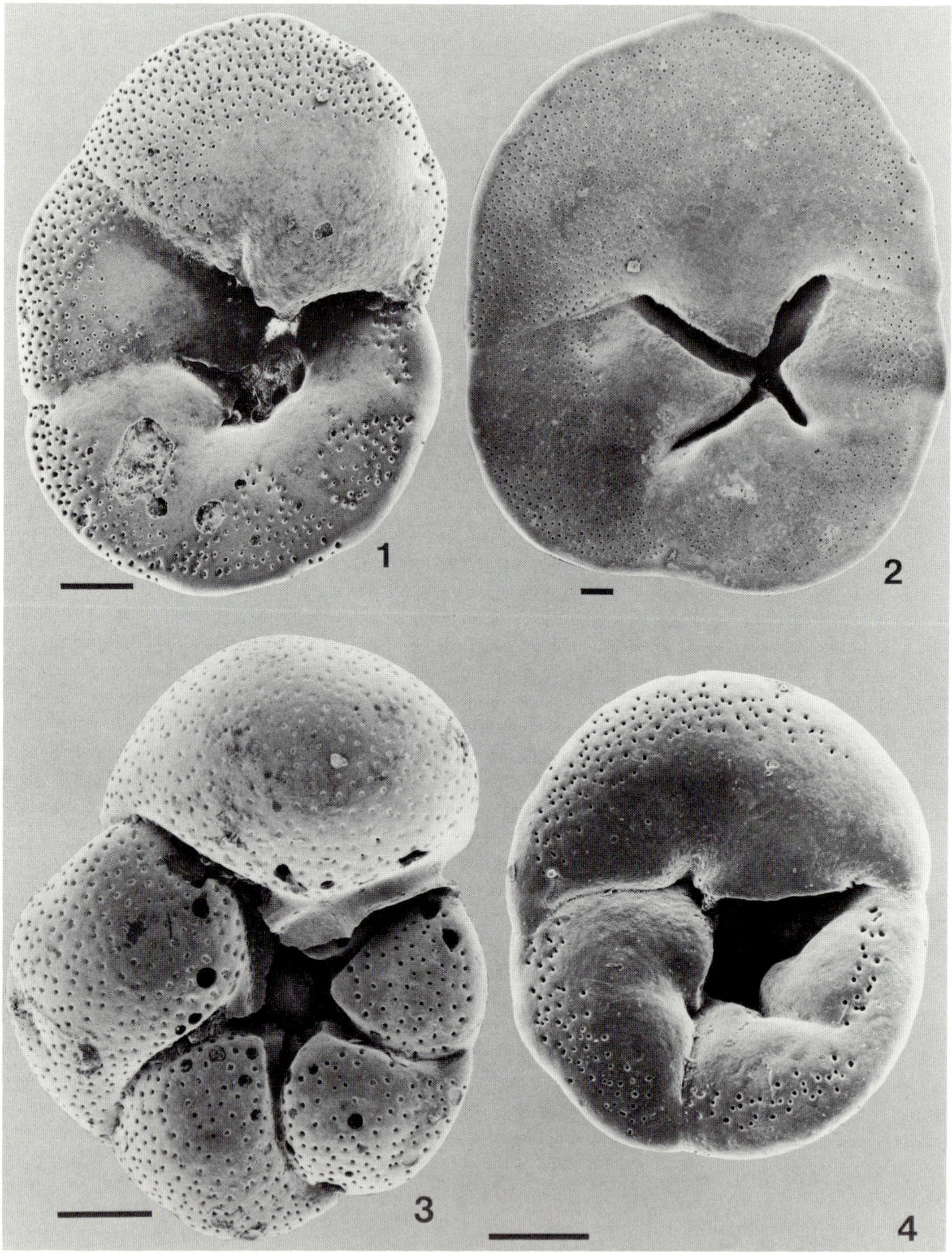

1. R. subauracana
2. R. floridensis
3. R. bahamaensis
4. R. concinna

ROSALINA

PLATE 42

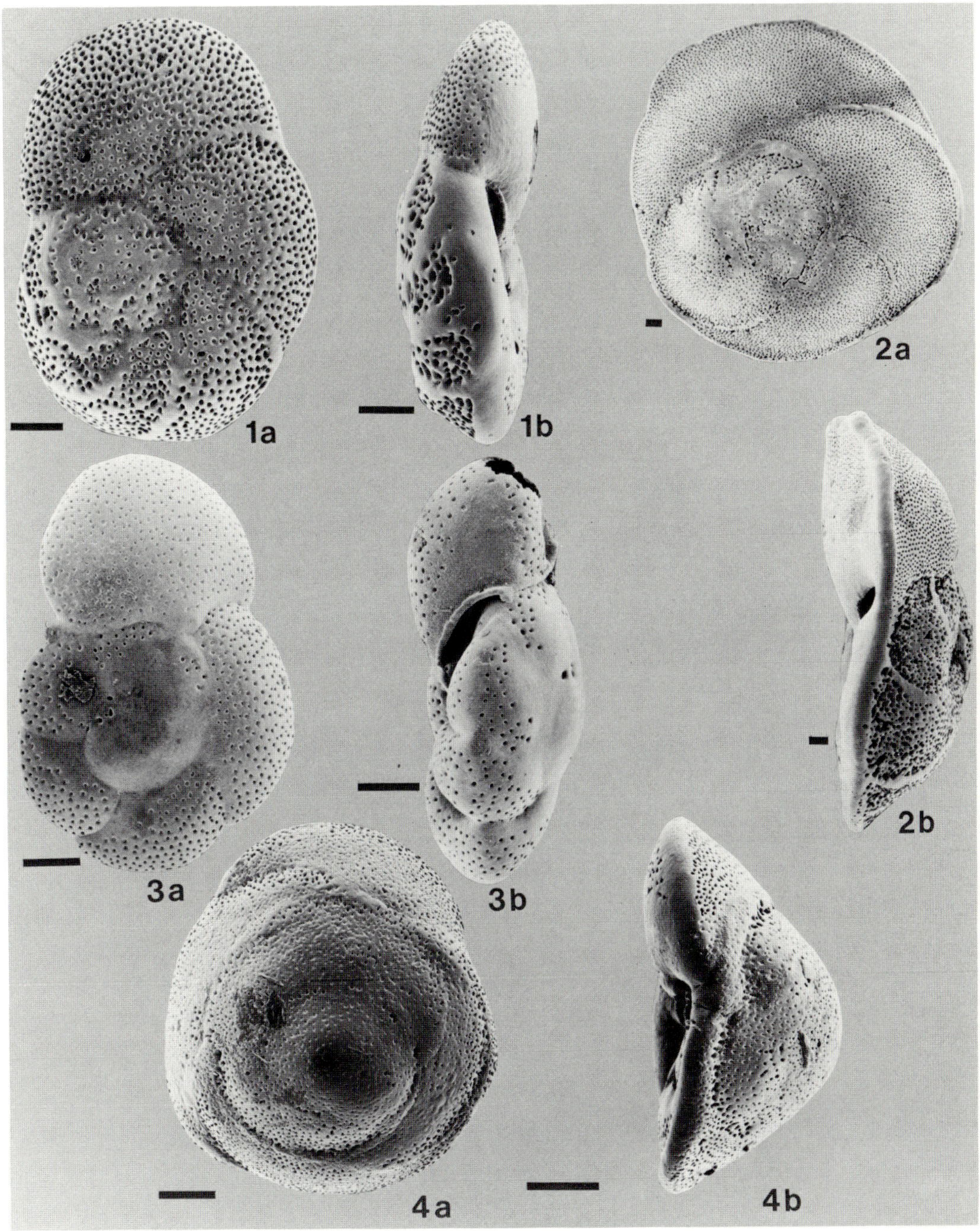

SHELF FACIES

PLATE 43

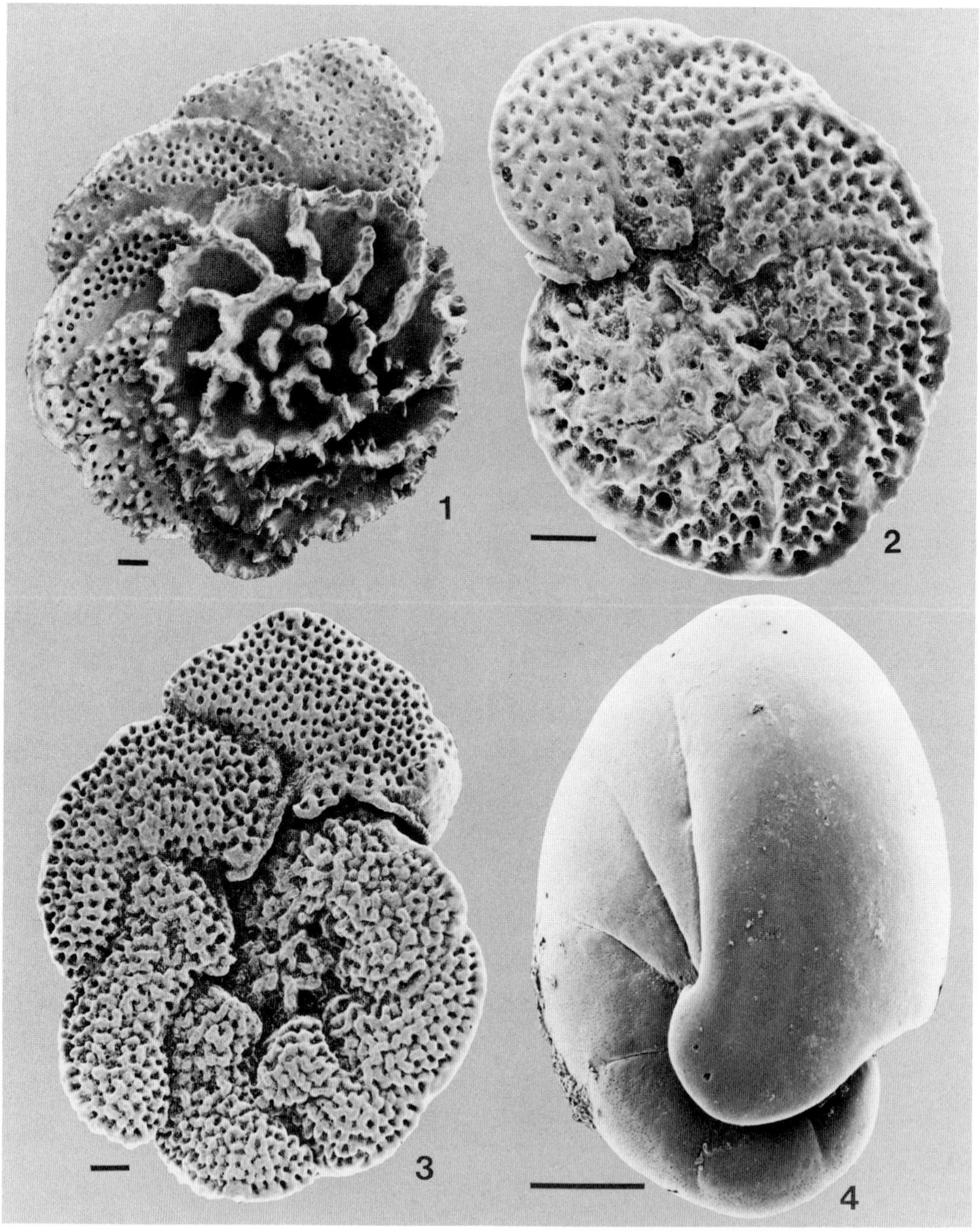

PLANULINA, NONIONELLA

PLATE 44

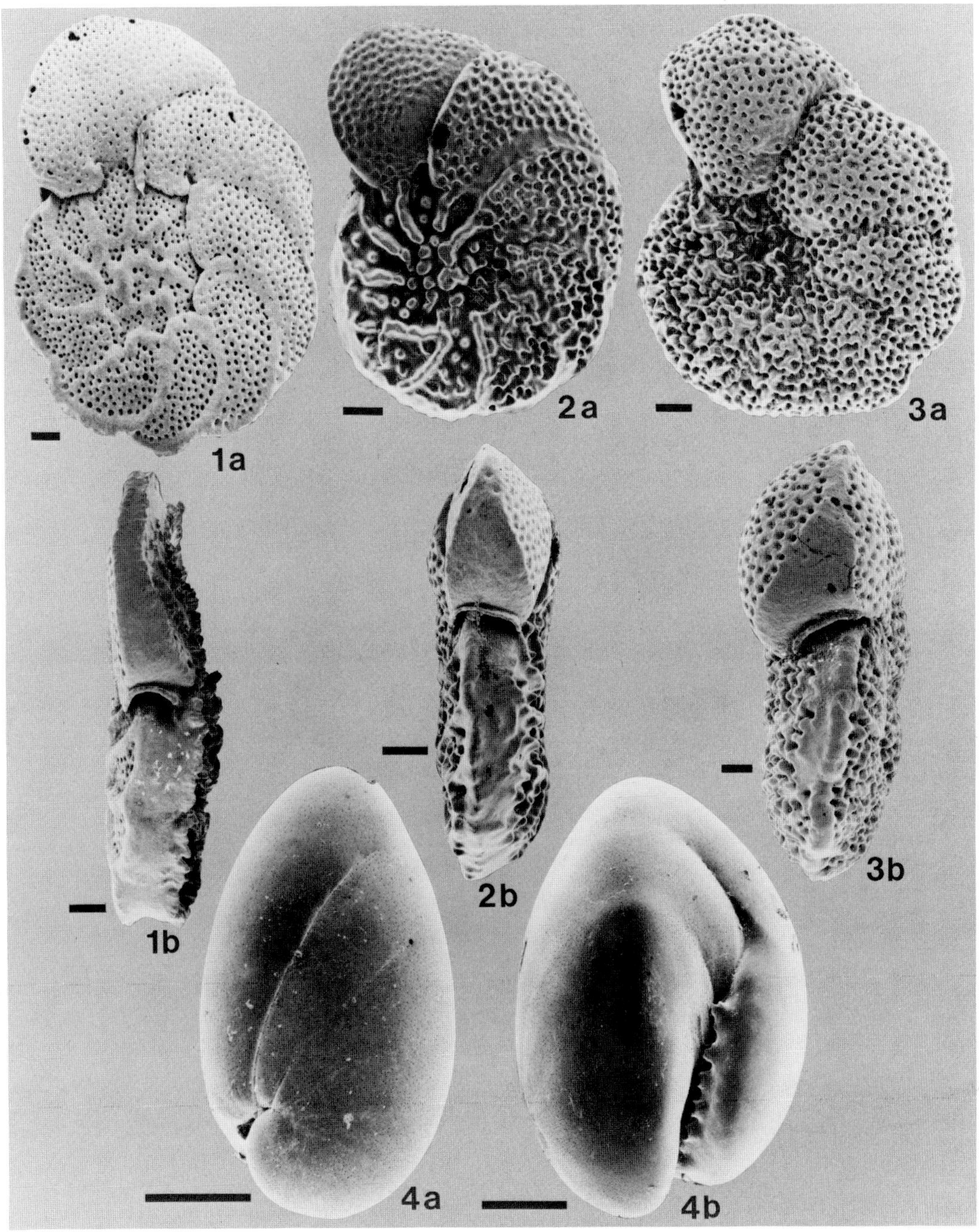

SHELF FACIES

PLATE 45

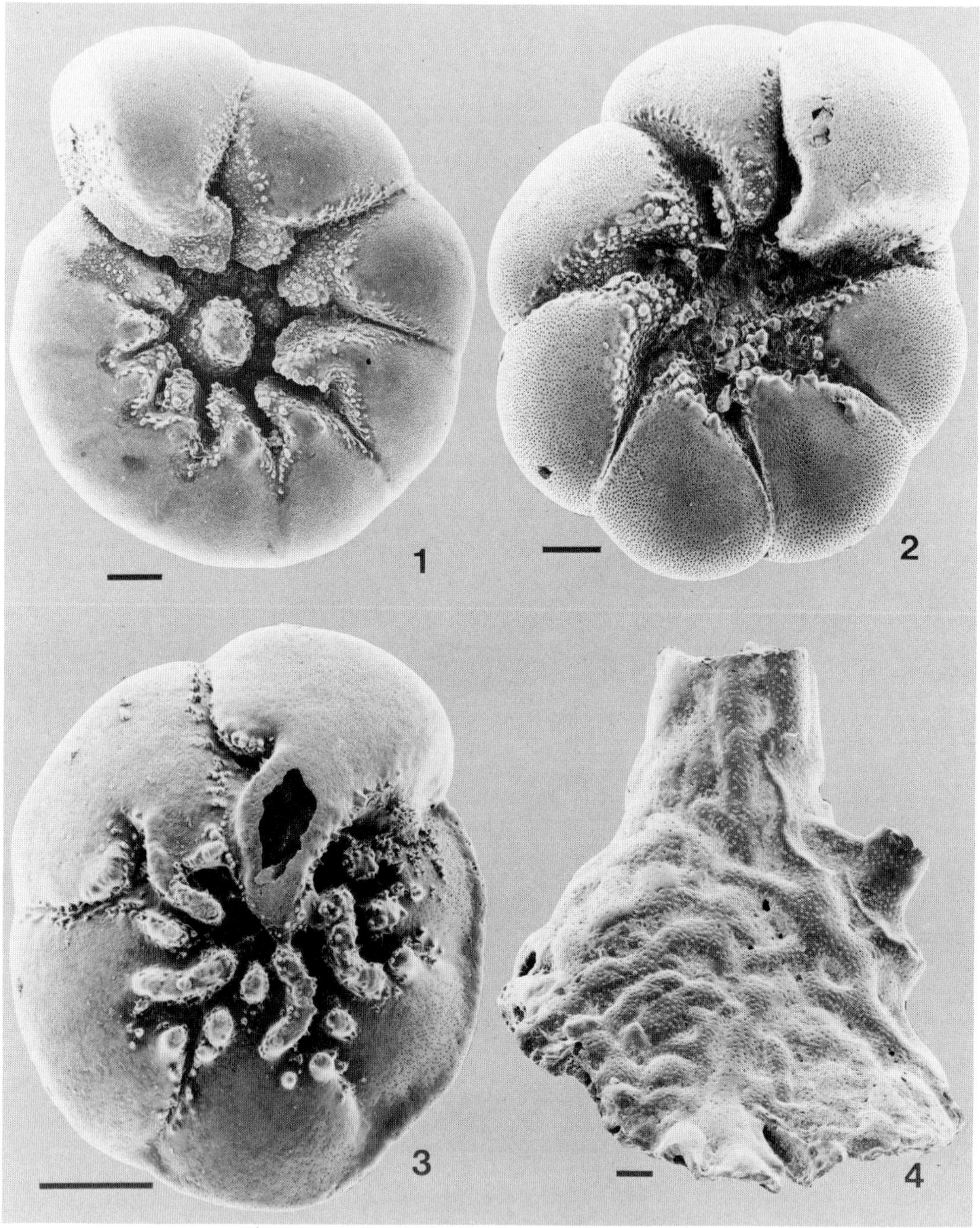

AMMONIA, HOMOTREMA

PLATE 46

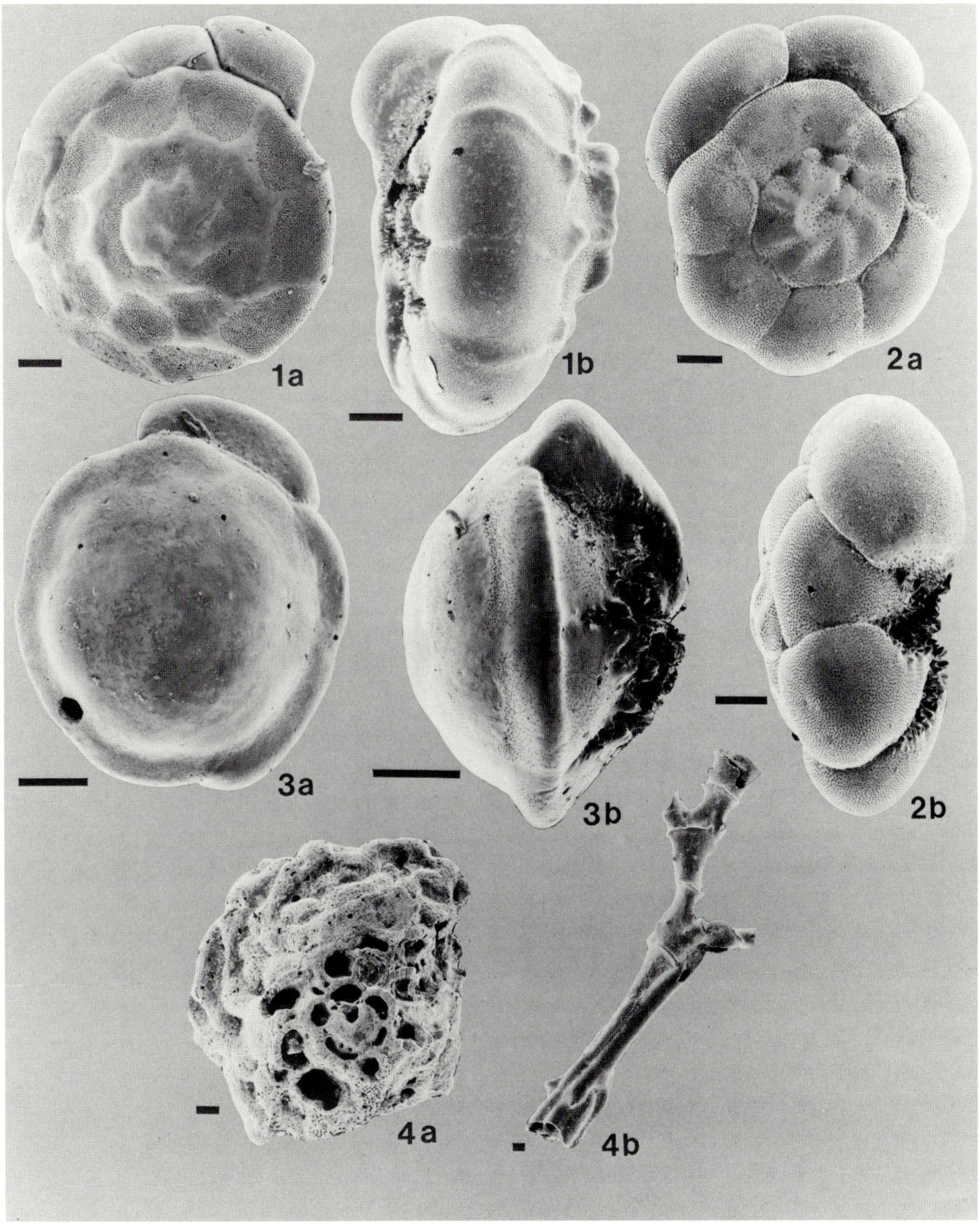

ESTUARY, SHELF, AND REEF FACIES

PLATE 47

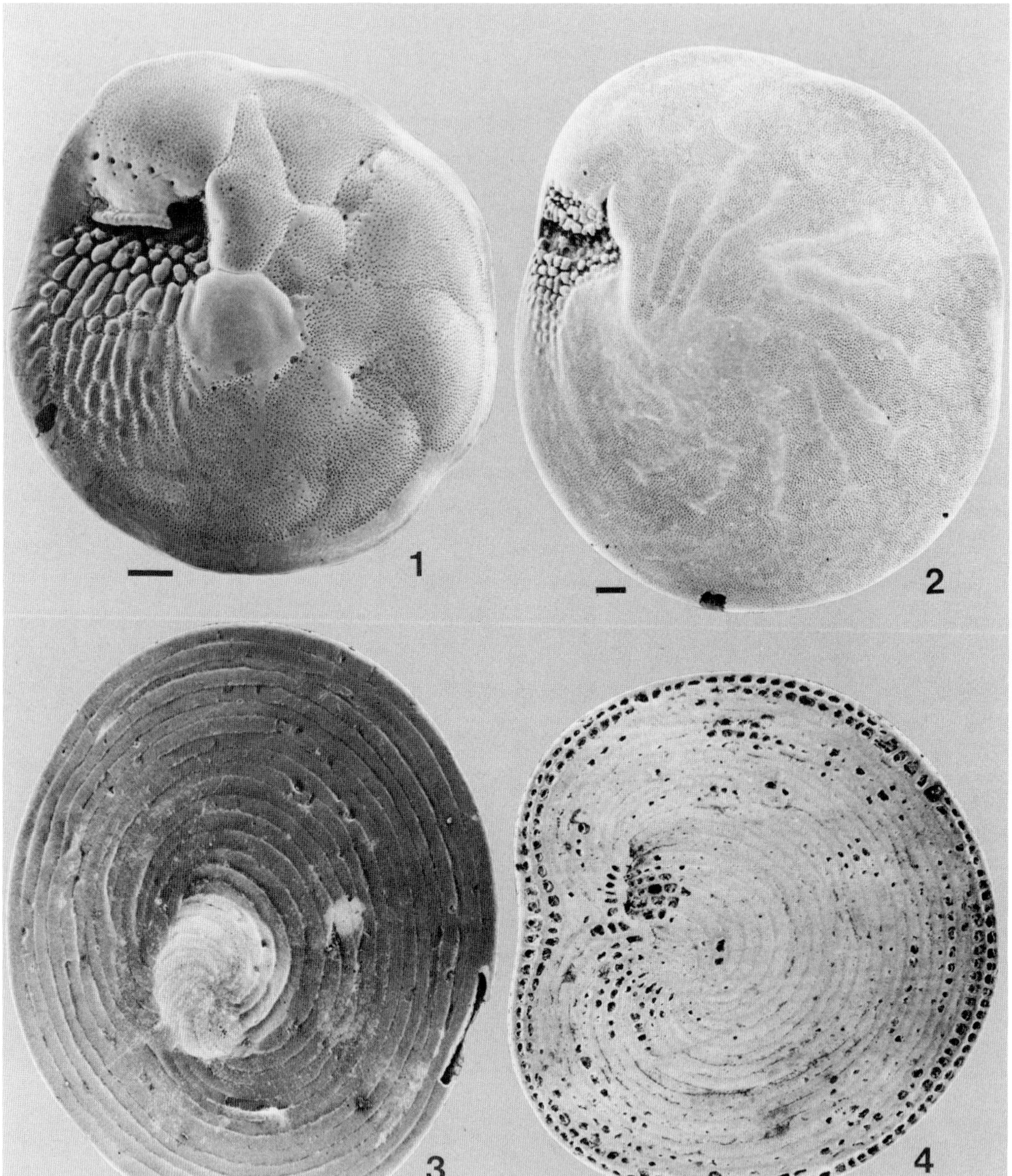

ASTERIGERINA, AMPHISTEGINA, PENEROPLIS, ARCHAIAS

PLATE 48

SHELF AND REEF FACIES

PLATE 49

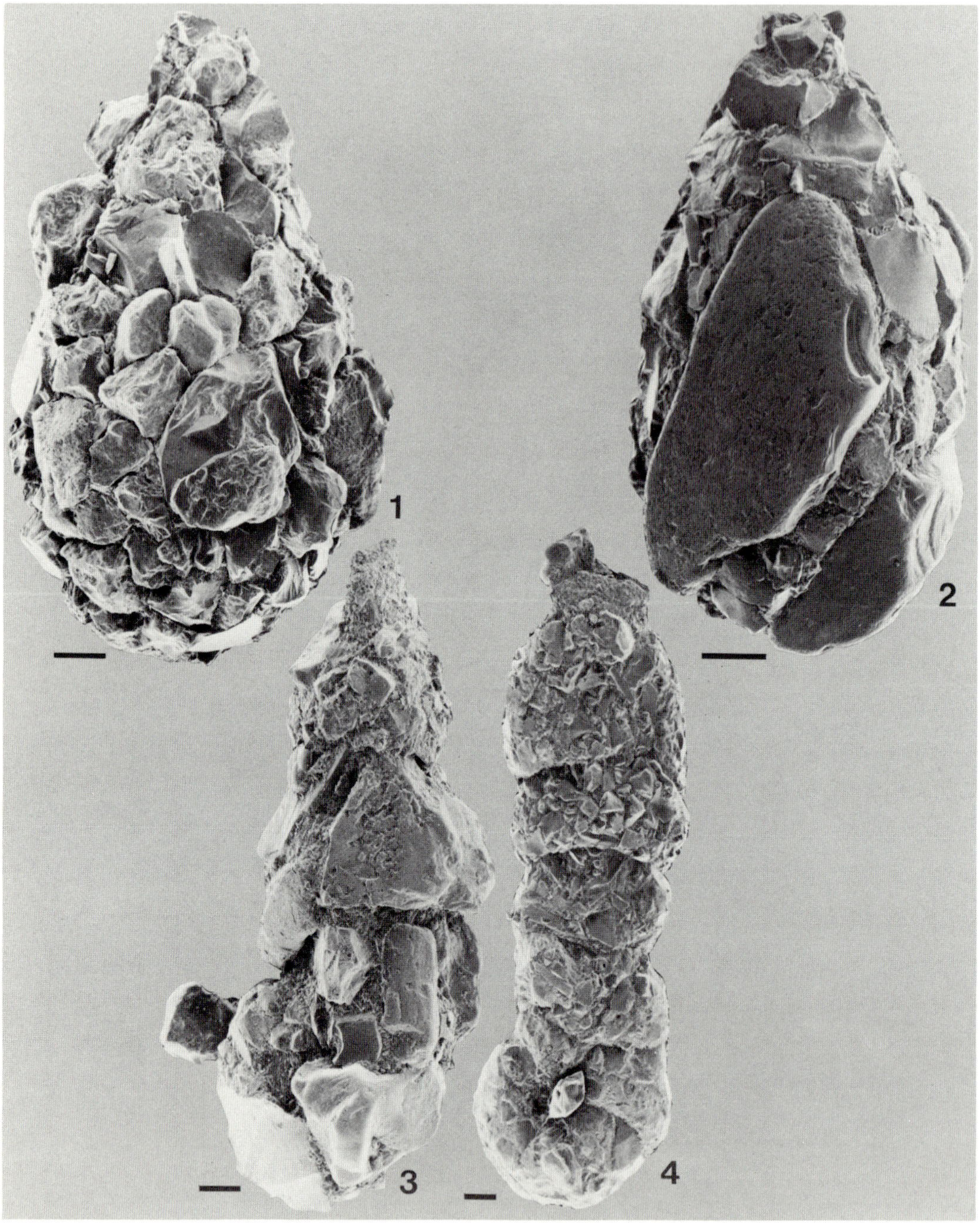

SACCAMMINA, NOURIA, AMMOBACULITES

PLATE 50

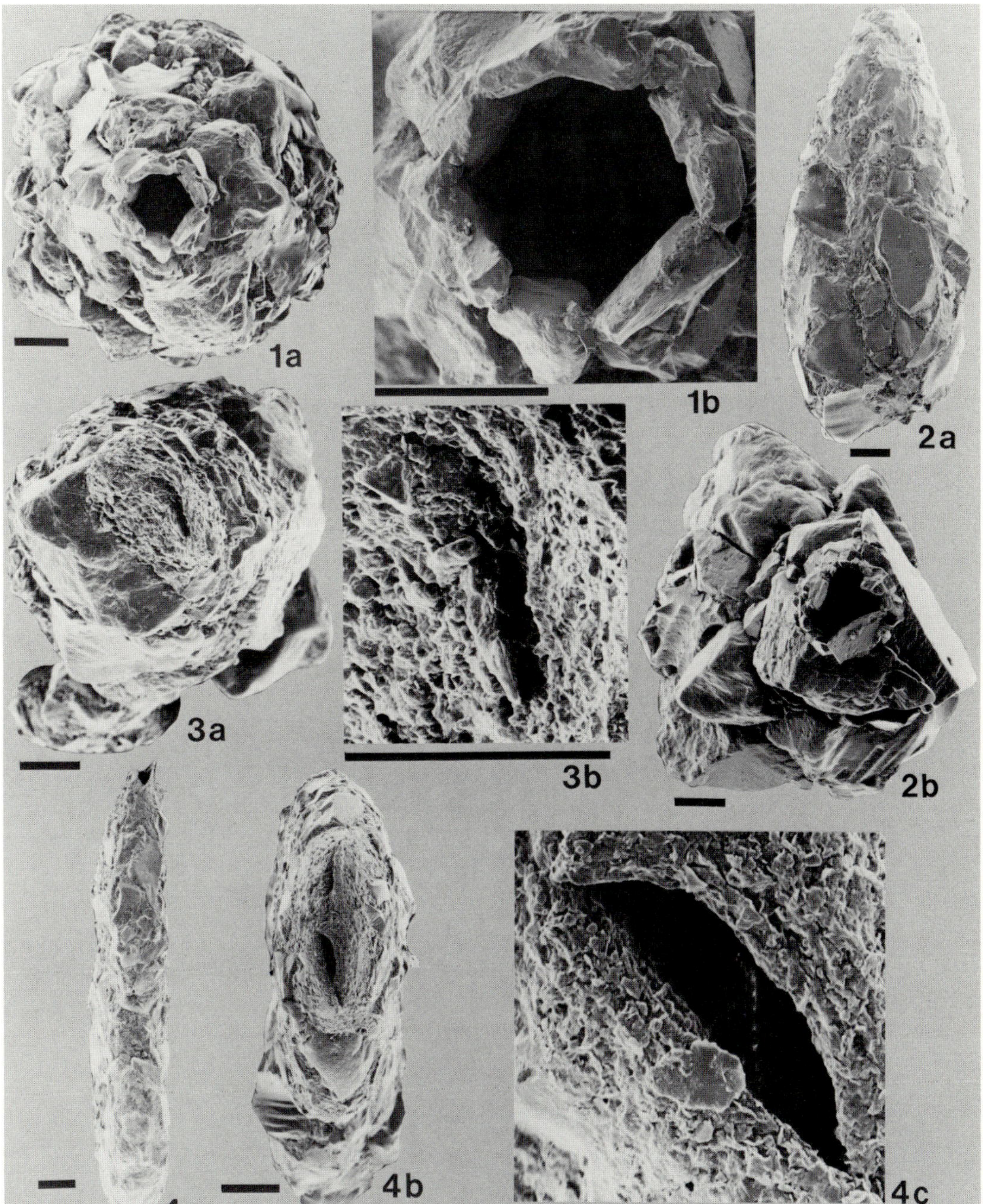

ESTUARY AND SHELF FACIES

PLATE 51

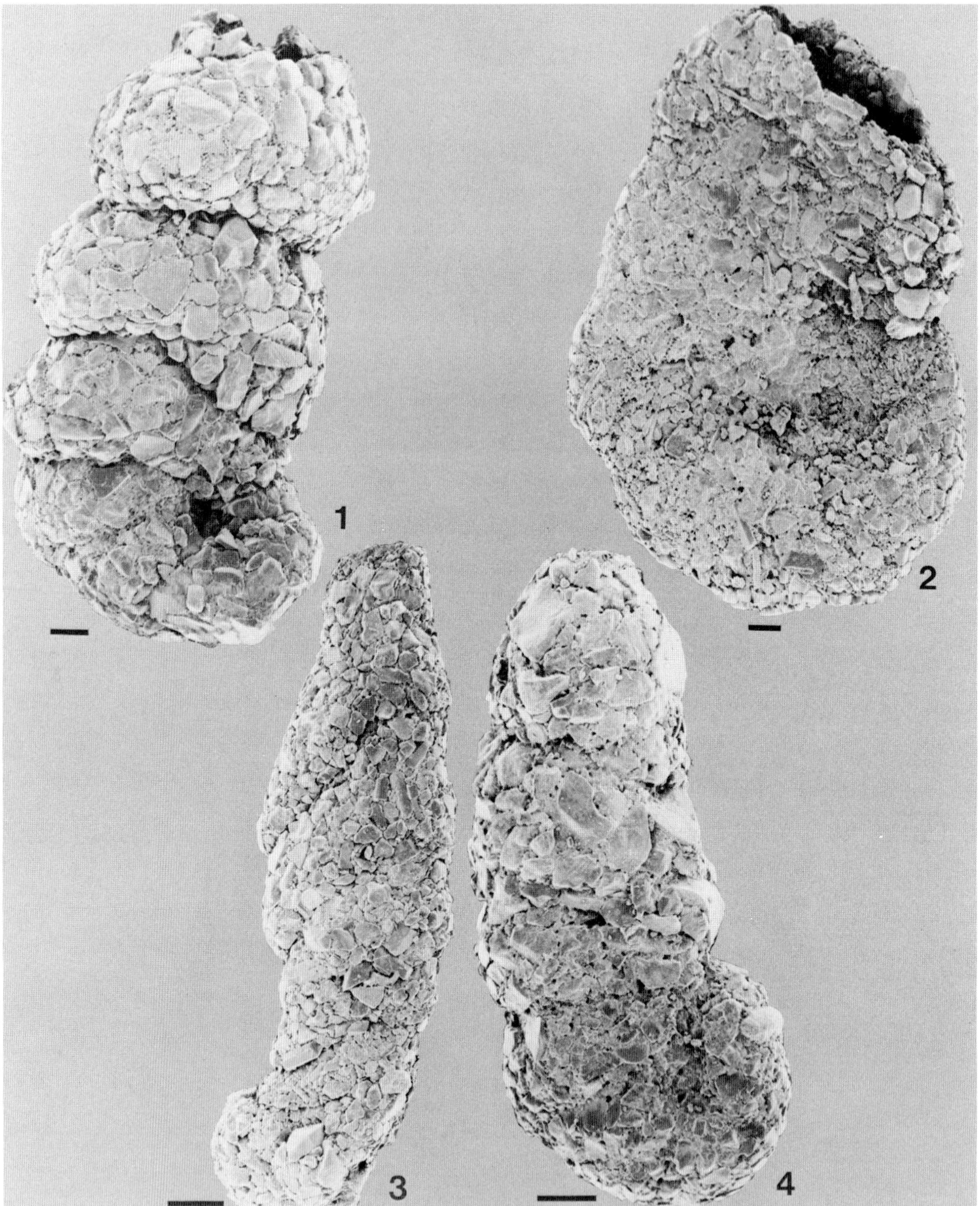

AMMOTIUM

PLATE 52

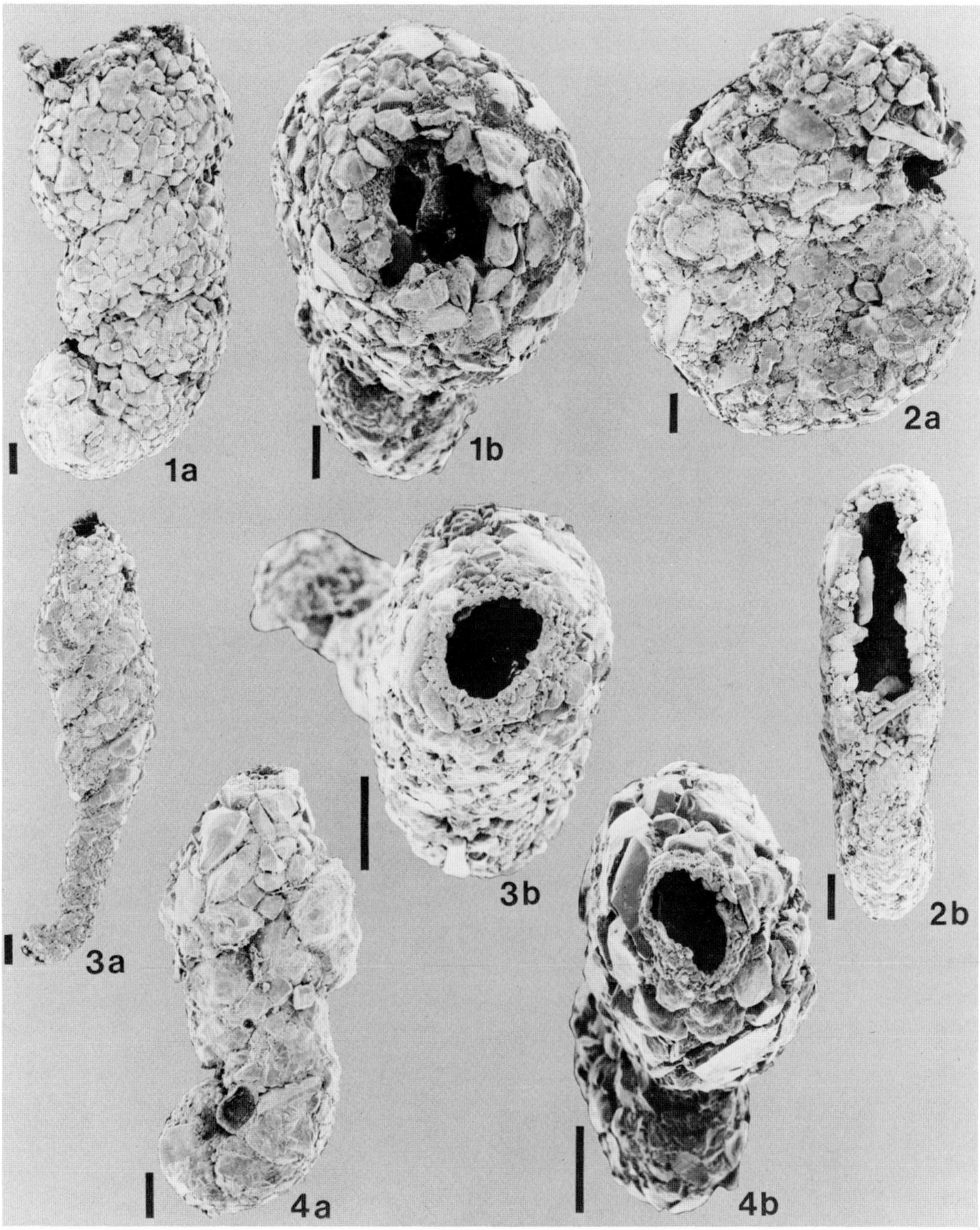

ESTUARY AND SHELF FACIES

PLATE 53

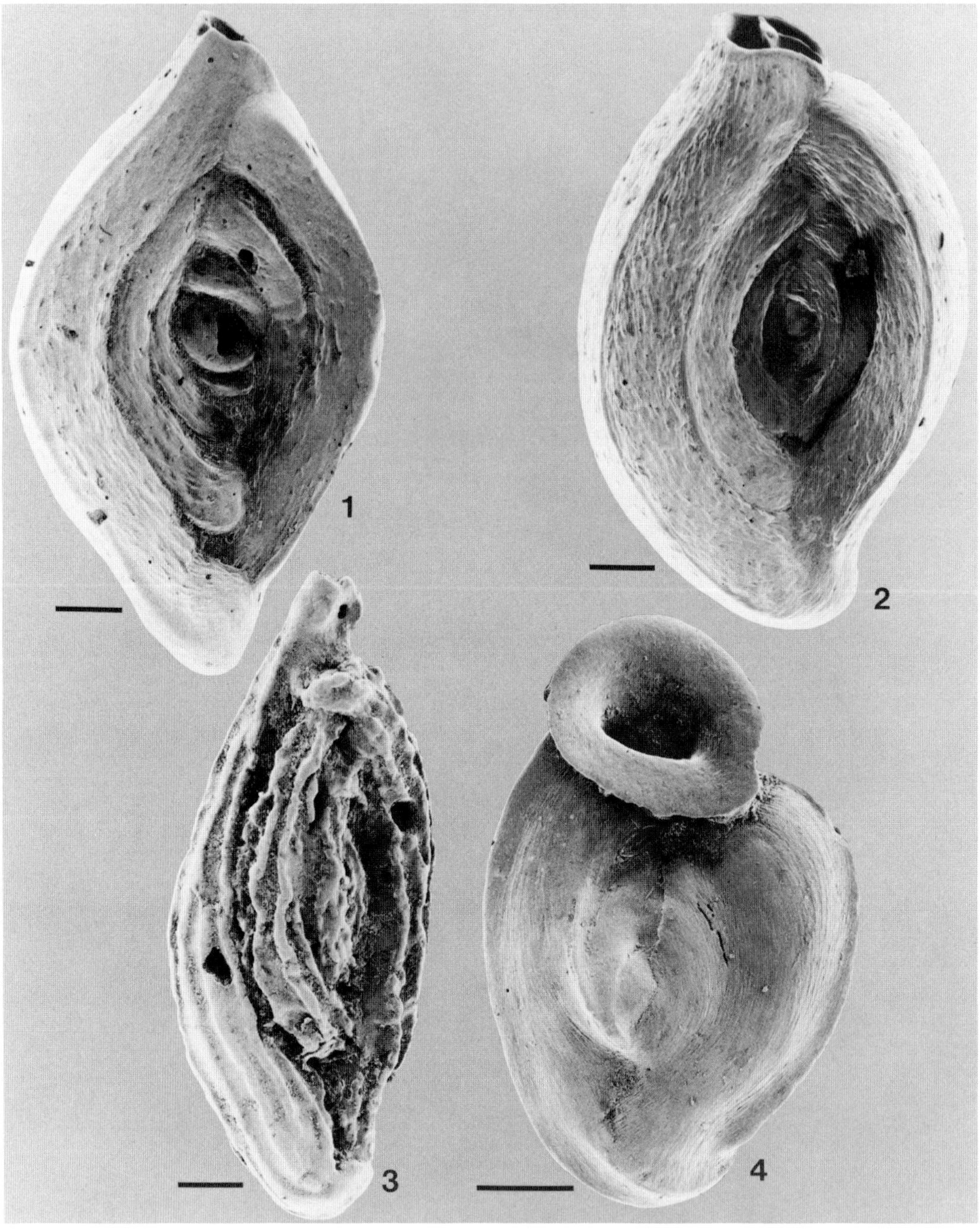

SPIROLOCULINA, WIESNERELLA

PLATE 54

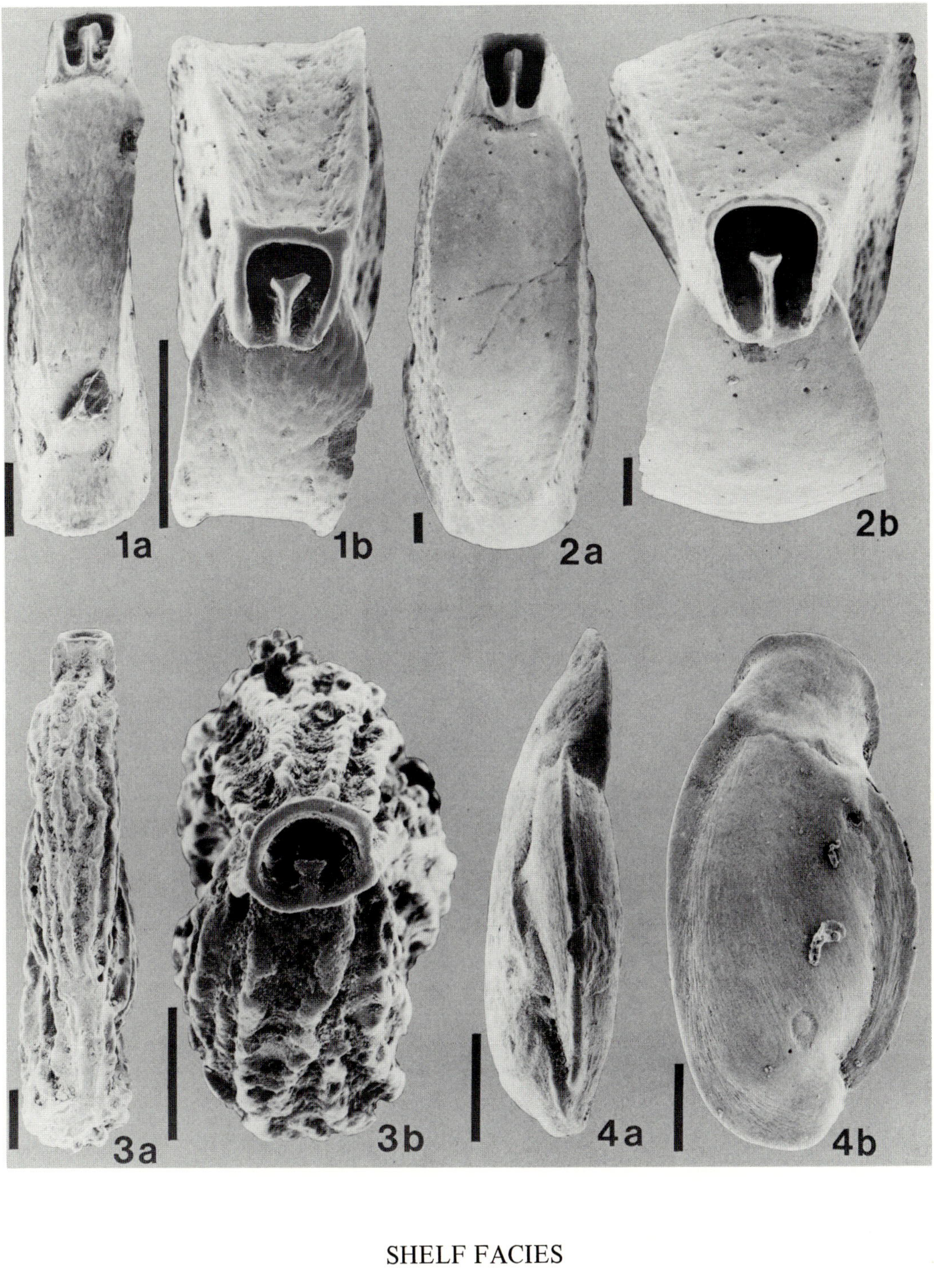

SHELF FACIES

PLATE 55

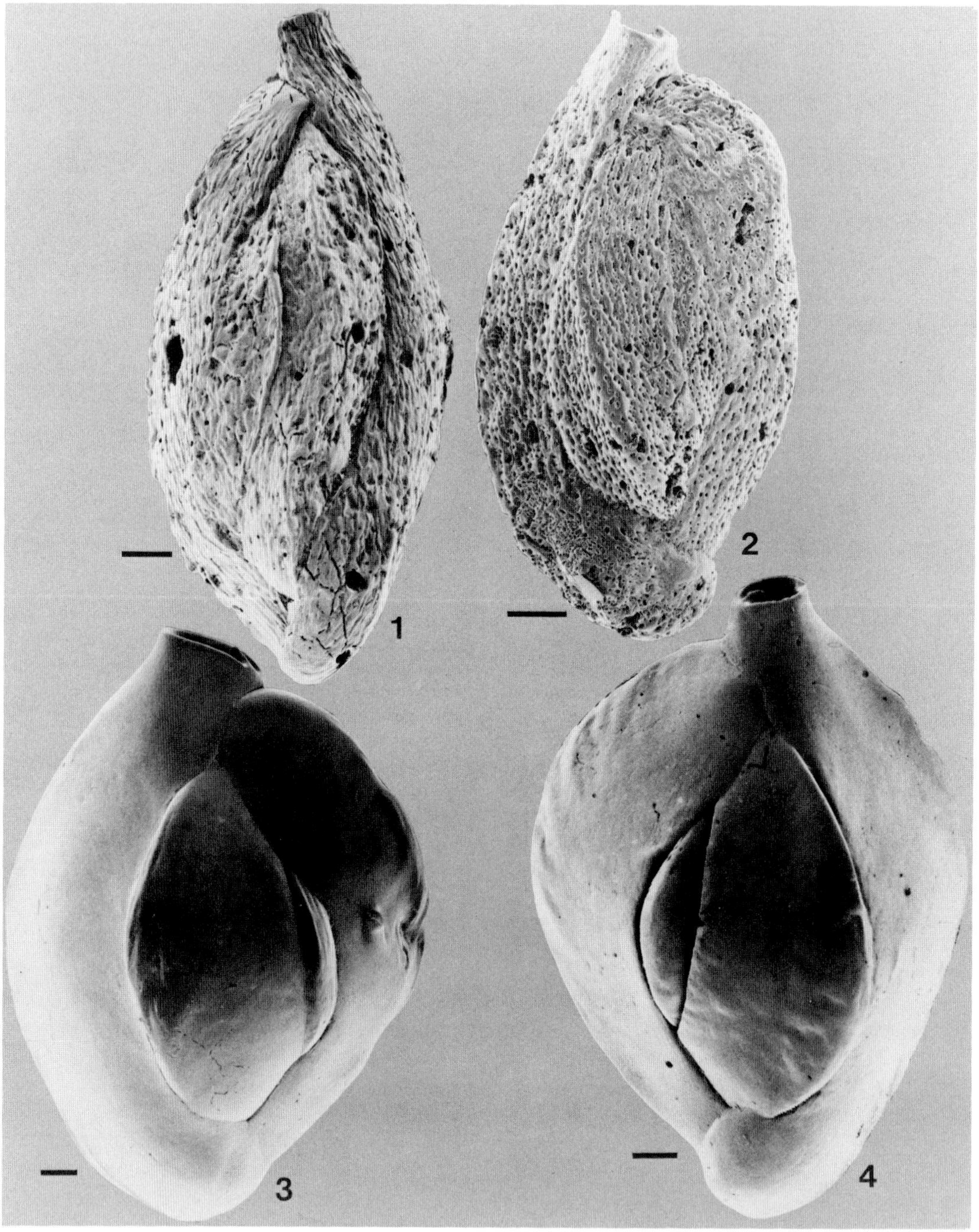

QUINQUELOCULINA

PLATE 56

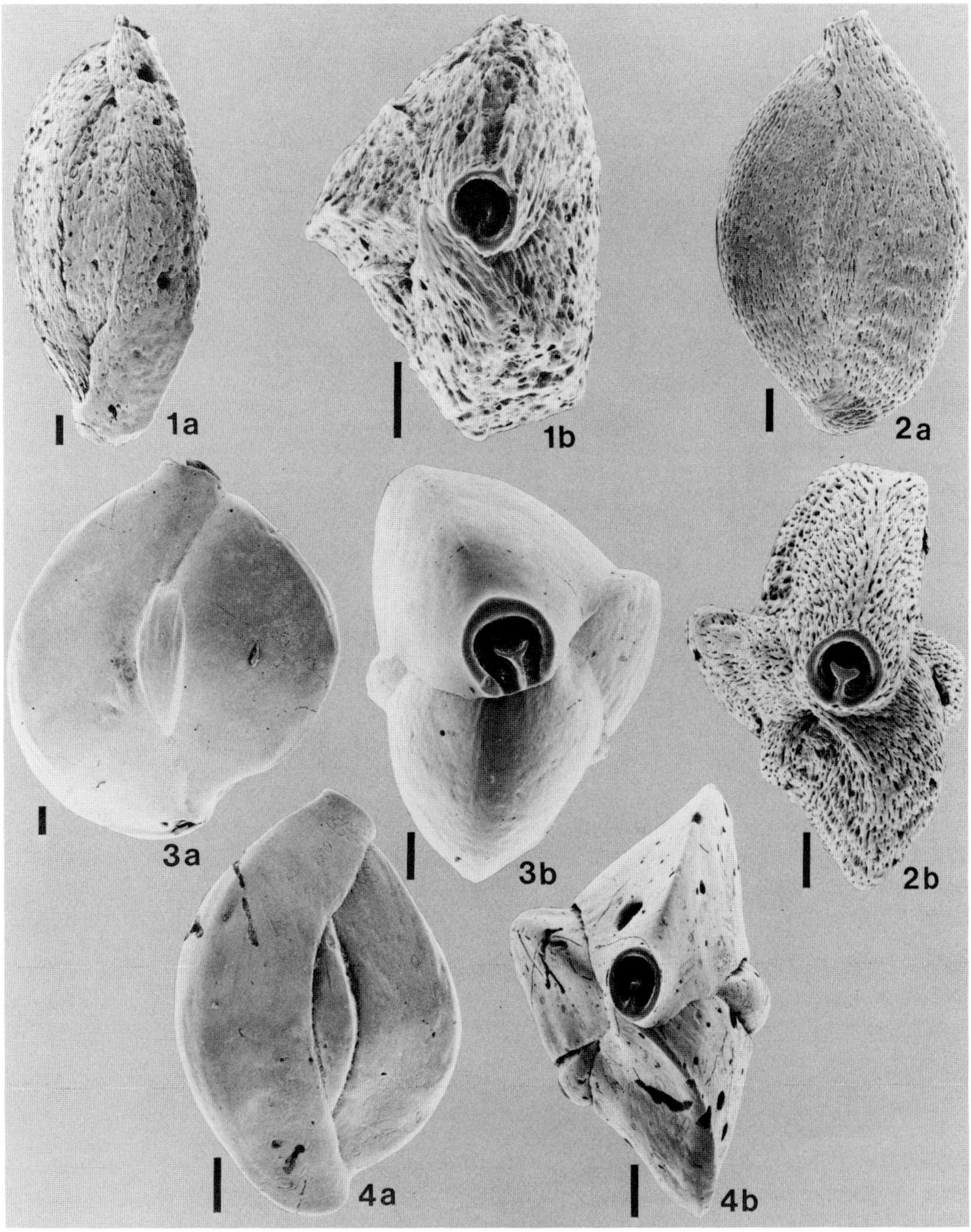

SHELF FACIES

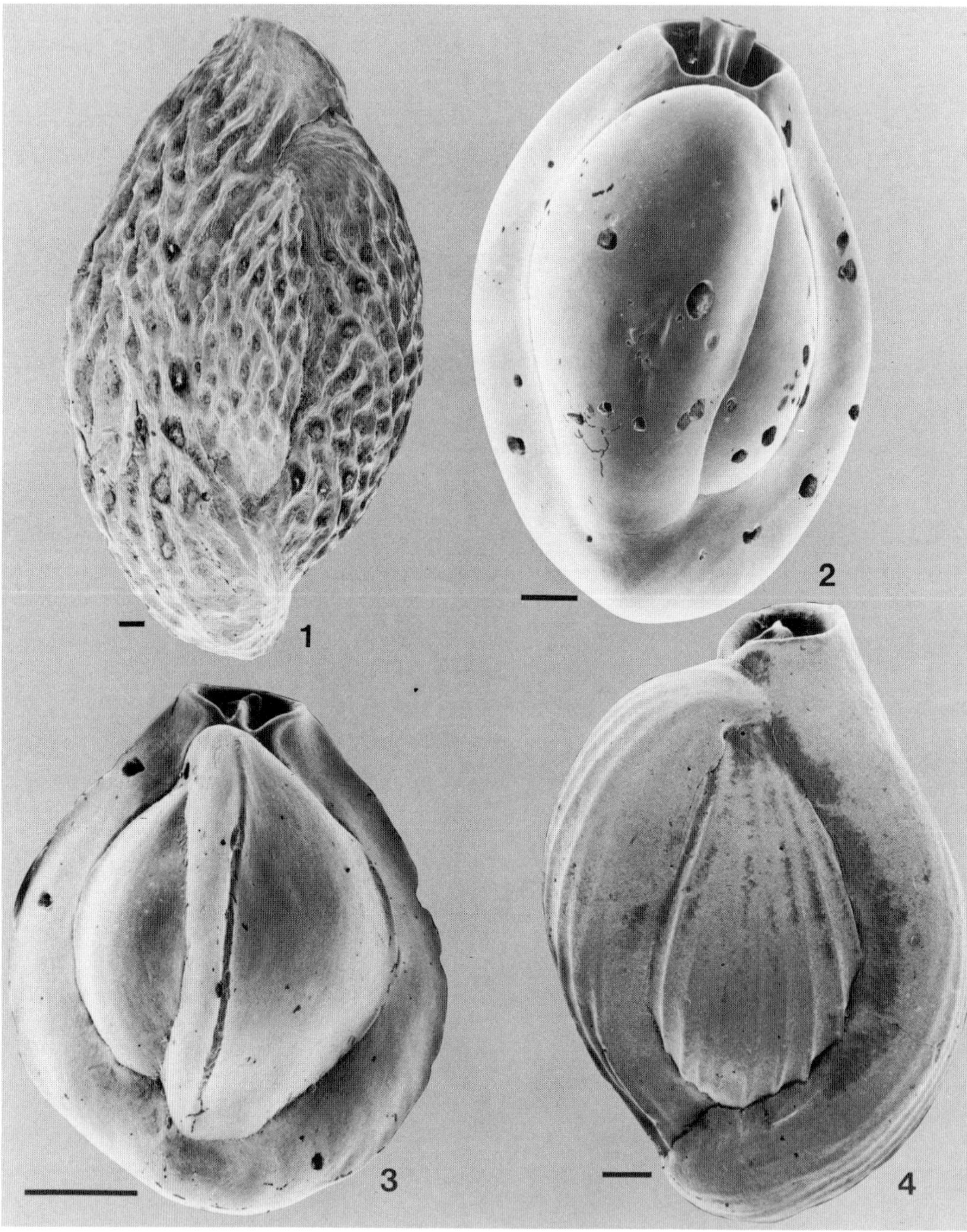

TRILOCULINA

PLATE 58

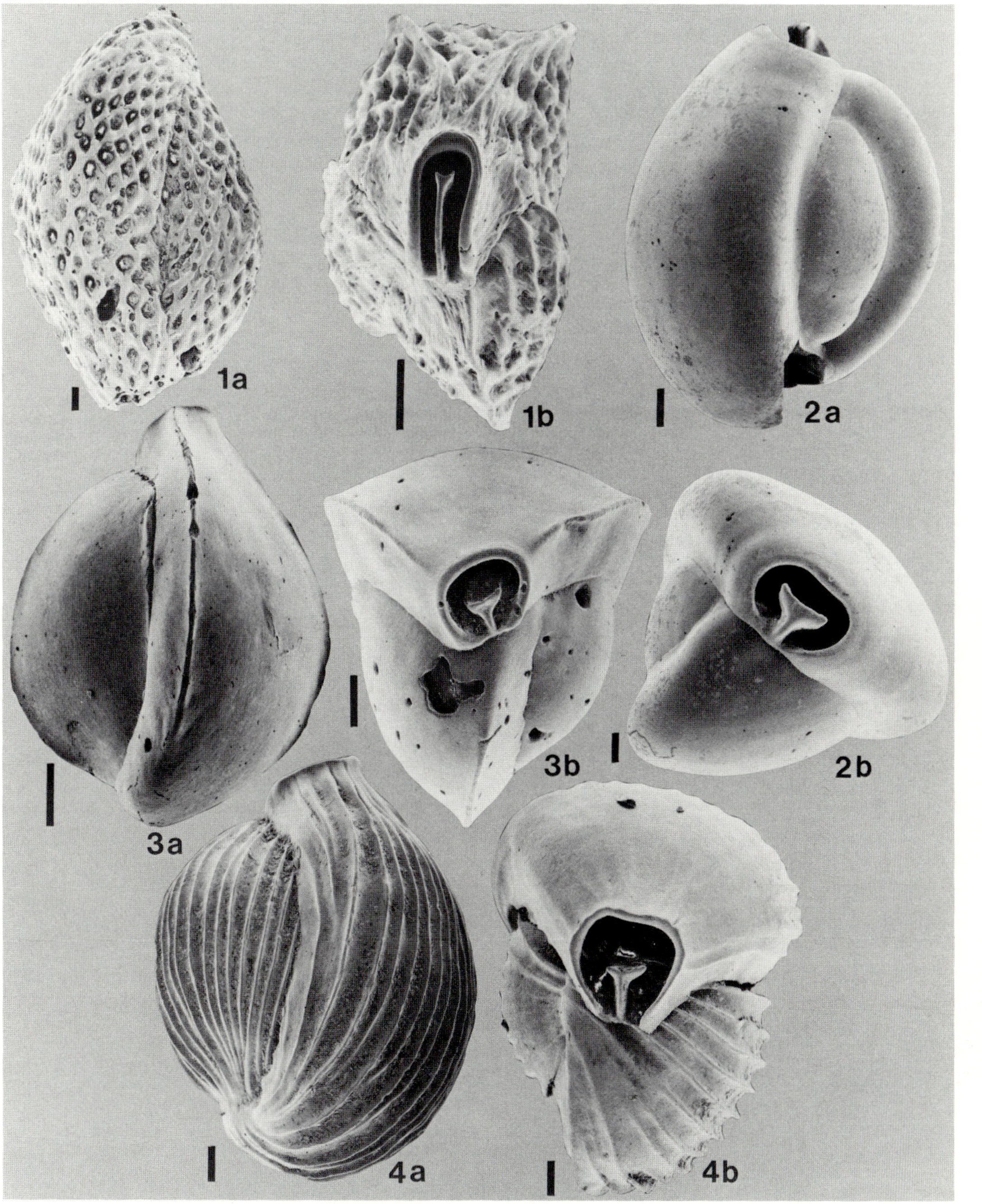

SHELF FACIES

PLATE 59

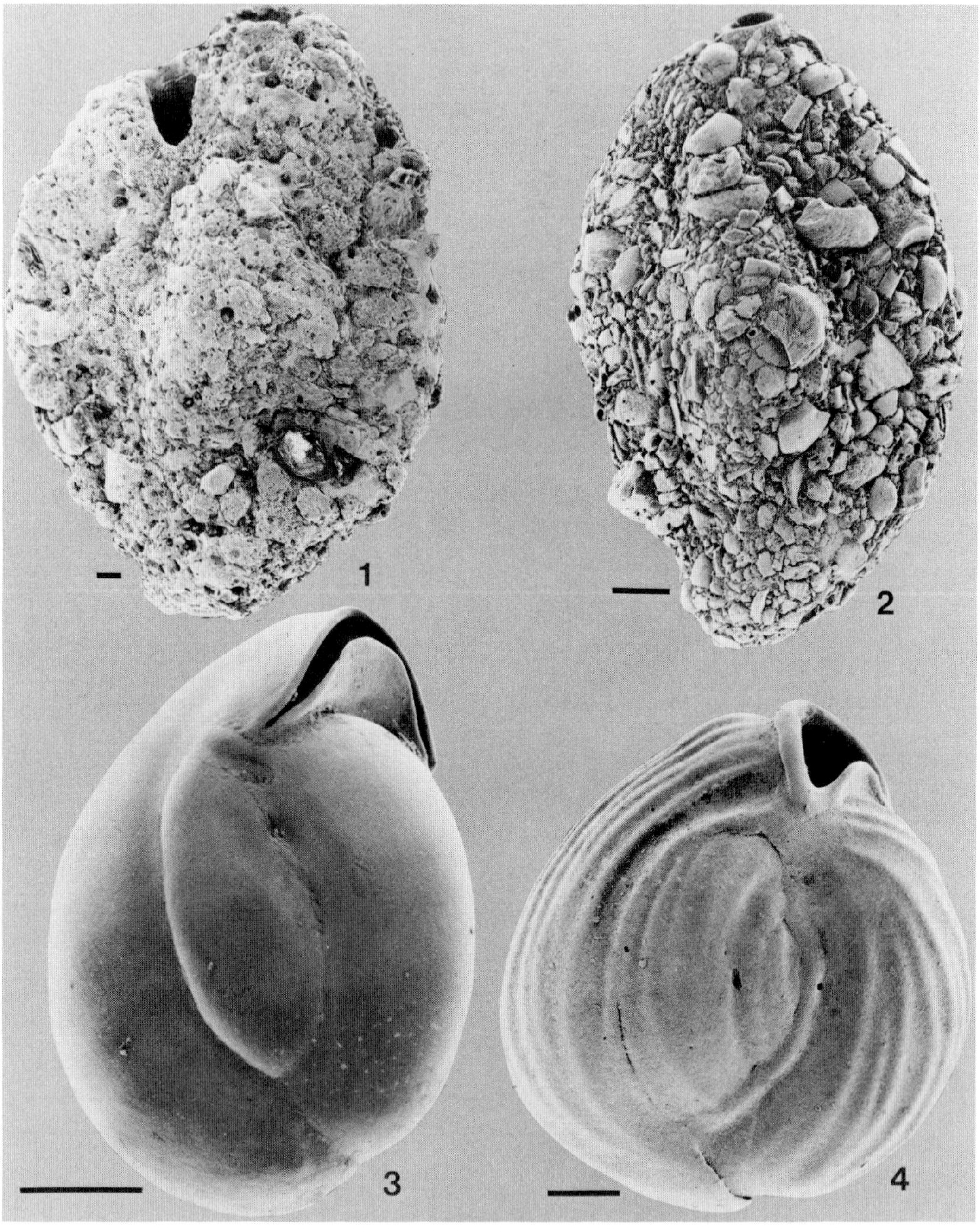

QUINQUELOCULINA, MILIOLINELLA

PLATE 60

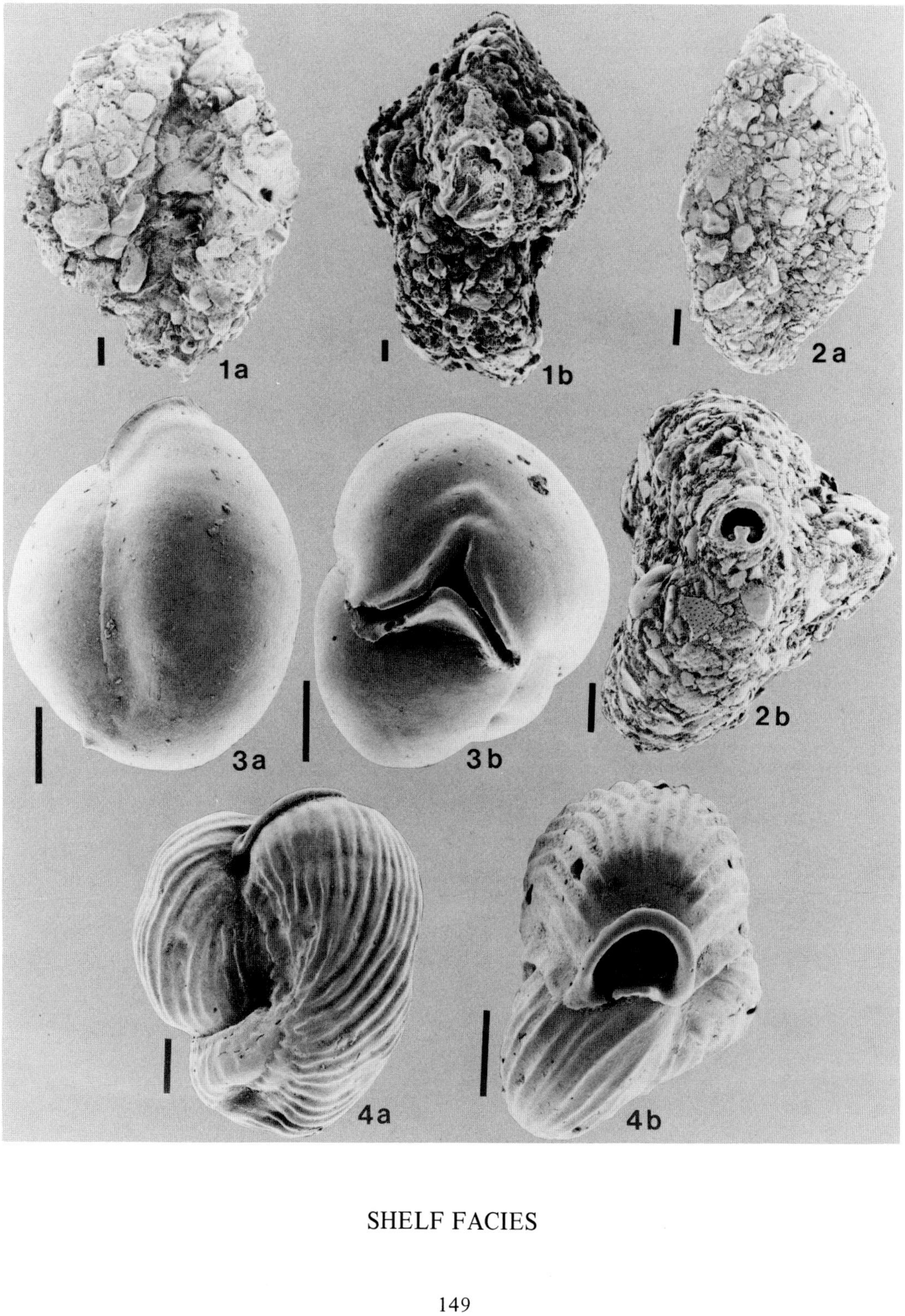

SHELF FACIES

PLATE 61

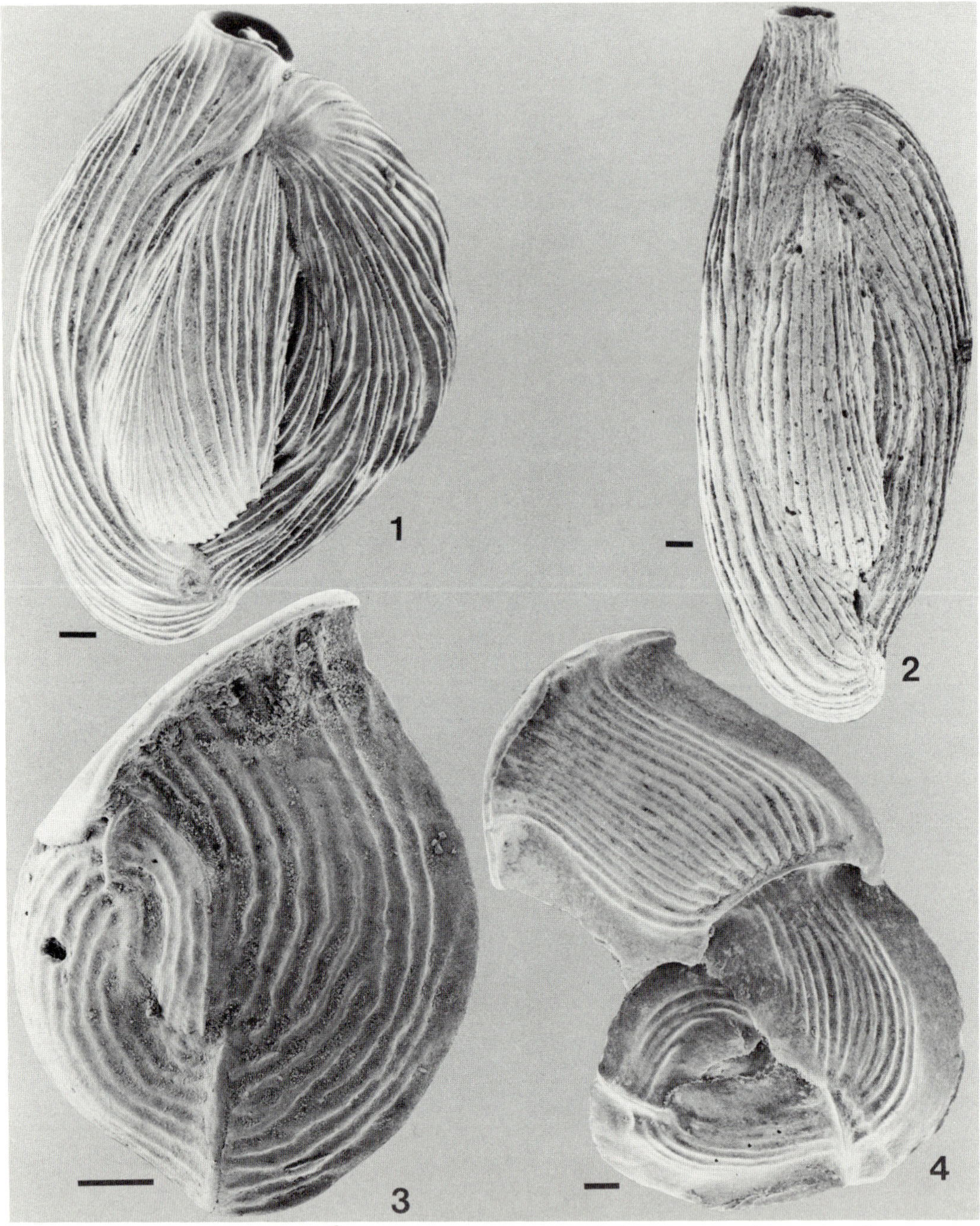

3. Articulina pacifica 4. Nodobaculariella cassis

QUINQUELOCULINA, ARTICULINA, NODOBACULARIELLA

PLATE 62

SHELF FACIES

PLATE 63

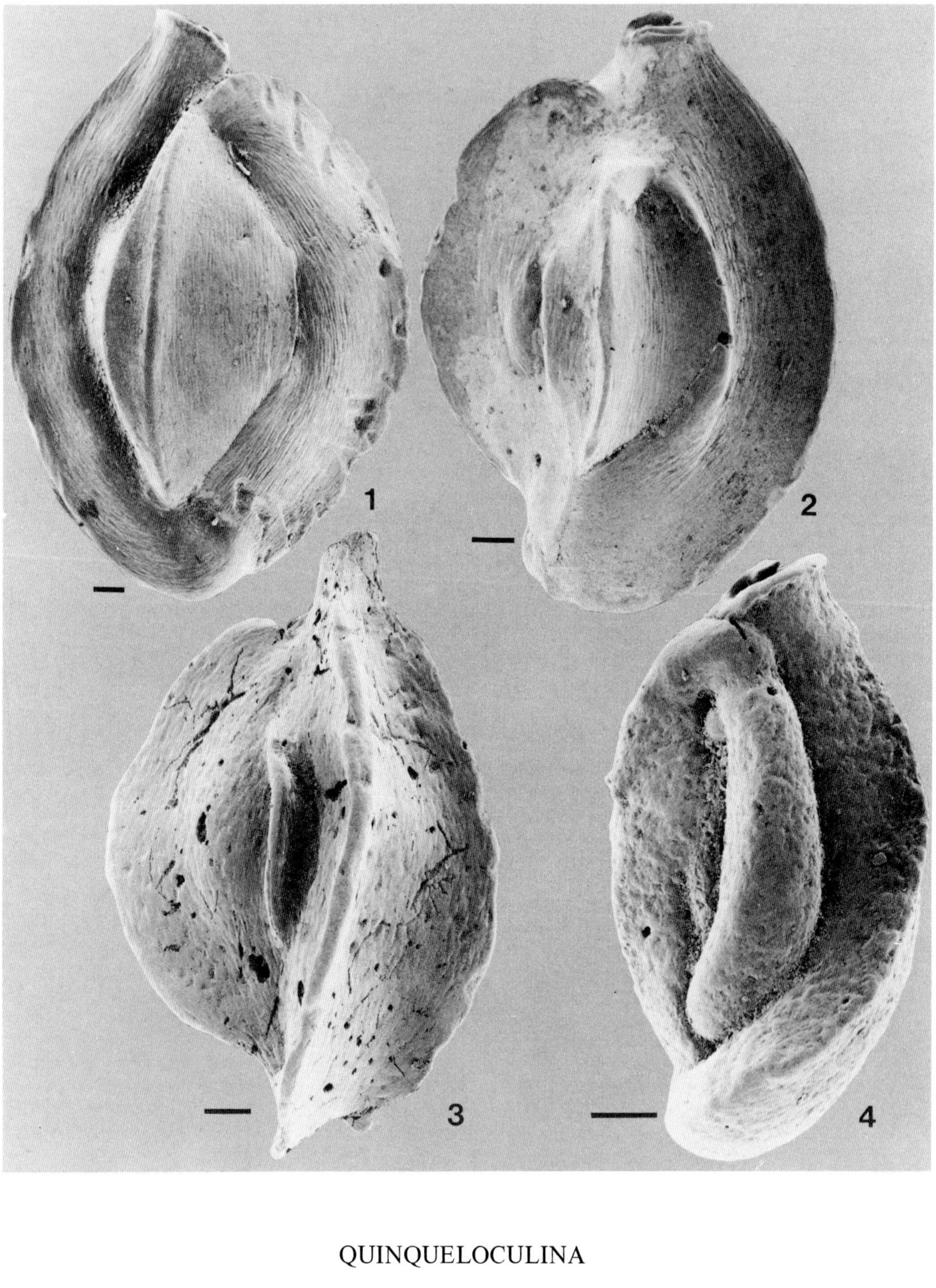

QUINQUELOCULINA

PLATE 64

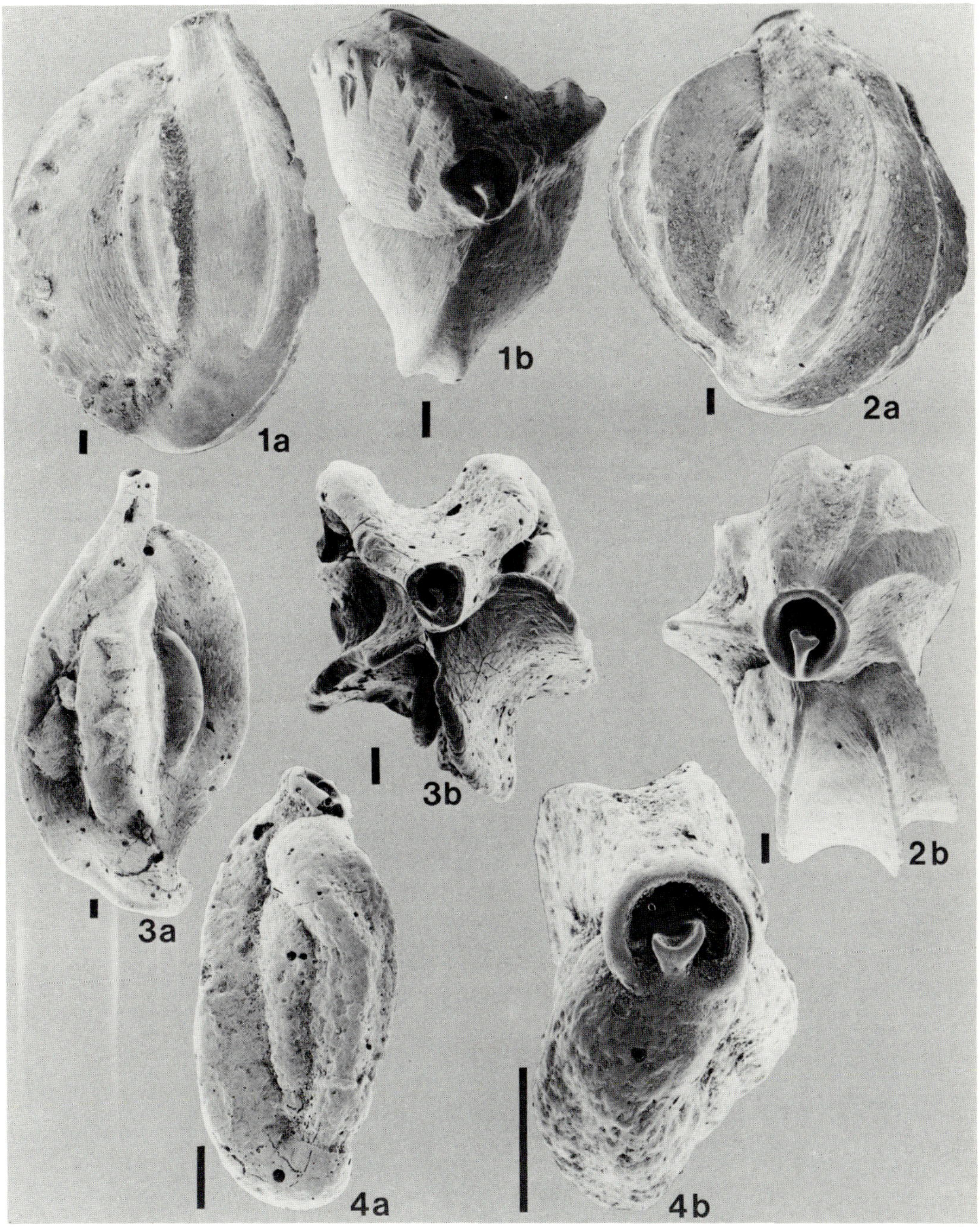

SHELF FACIES

References

Ager, D. V. 1963. *Principles of paleoecology*. New York: McGraw-Hill, 371 p.

Anepohl, J. K. 1976. Seasonal distribution of living benthonic foraminifera of the south Texas outer continental shelf. Master's thesis, University of Texas.

Ashton, J. H., and A. J. Rowell. 1975. Environmental stability and species proliferation in Late Cambrian trilobite faunas: a test of the niche-variation hypothesis. *Paleobiology* 1:161-174.

Ayala-Castañares, A. 1963. Sistematica y distribucion de los Foraminiferos Receientes de la Laguna de Terminos, Campeche, Mexico. *Univ. Nal. Auton. Mexico Inst. Geol. Bol. 67*, pt. 3, 130 p.

Ballard, R. D., and E. Uchupi. 1970. Morphology and Quaternary history of the continental shelf of the Gulf Coast of the United States. *Bull. Marine Sci*. 20:547-559.

Bambach, R. K. 1977. Species richness in marine benthic habitats through the Phanerozoic. *Paleobiology* 3:152-167.

Bandy, O. L. 1954. Distribution of some shallow-water foraminifera in the Gulf of Mexico. *U.S. Geol. Survey, Prof. Paper 254-F*, p. 125-141.

Bandy, O. L. 1956. Ecology of foraminifera in northeastern Gulf of Mexico. *U.S. Geol. Survey, Prof. Paper 274-G*, p. 179-204.

Belanger, P. E., and S. S. Streeter. 1979. Distribution and ecology of benthic foraminifera in the Norwegian-Greenland Sea. *Marine Micropaleont*. 5:401-428.

Benda, W. R., and H. S. Puri. 1962. The distribution of foraminifera and ostracoda off the Gulf Coast of the Cape Romano area, Florida. *Gulf Coast Assoc. Geol. Socs., Trans*. 12:303-342.

Bergantino, R. N. 1971. Submarine regional geomorphology of the Gulf of Mexico. *Geol. Soc. America Bull*. 82:741-752.

Berggren, W. A., and B. U. Haq. 1976. The Andalusian Stage (Late Miocene): biostratigraphy, biochronology, and paleoecology. *Palaeogeogr., Palaeoclimatol., Palaeoecol*. 20: 67-129.

Berggren, W. A.; R. H. Benson; B. U. Haq; W. R. Reidel; A. Sanfilippo: H. J. Schrader: and R. C. Tjalsma. 1976. The El Cuervo section (Andalusia Plain): micropaleontologic anatomy of an early late Miocene lower bathyal deposit. *Marine Micropaleont*. 1:195-247.

Bernstein, B. B.; R. R. Hessler; R. Smith; and P. A. Jumars. 1978. Spatial dispersion of benthic foraminifera in the abyssal central North Pacific. *Limnol. Oceanogr*. 23:401-416.

Bock, W. D. 1961. The benthonic foraminifera of southwestern Florida Bay. Master's thesis, University of Wisconsin.

Bock, W. D. 1967. Monthly variation in the foraminiferal biofacies on *Thallasia* and sediment in the Big Pine Key area, Florida. Ph.D. thesis, University of Miami.

Bock, W. D.; W. W. Hay; J. I. Jones; G. W. Lynts; S. L. Smith; and R. C. Wright. 1971. A symposium of Recent South Florida foraminifera. *Miami Geol. Soc. Mem. 1*, 245 p.

Bock, W. D. 1976. Distribution and significance of foraminifera in the MAFLA area. In C. T. Schafer and B. R. Pelletier, eds., *First international symposium on benthonic foraminifera of continental margins, Part A, Ecology and Biology*. Maritime Seds., Spec. Publ. No. 1. p. 221-237.

Bouma, A. H. 1968. Distribution of minor structures in Gulf of Mexico sediments. *Gulf Coast Assoc. Geol. Socs., Trans*. 18:26-33.

Bouma, A. H. 1972. Distribution of sediments and sedimentary structures in the Gulf of Mexico. In R. Rezak and V. J. Henry, eds., *Contributions on the geological and geophysical oceanography of the Gulf of Mexico*. Texas A&M Univ. Oceanogr. Stud. Ser. 3:35-65.

Bouma, A. H. 1973. Leveed-channel deposits, turbidites, and contourites in deeper part of Gulf of Mexico. *Gulf Coast Assoc. Geol. Socs., Trans*. 23:368-376.

Bouma, A. H., ed. 1976. *Shell dredging and its influence on Gulf Coast environments*. Houston: Gulf Publ., 454 p.

Bouma, A. H.; L. B. Smith; B. R. Sidner; and T. R. McKee. 1975. *Submarine geomorphology and sedimentation patterns of the Gyre Intraslope Basin, northwest Gulf of Mexico*. Texas A&M Research Found. Tech. Rept. 75-9-T, 163 p.

Bouma, A. H.; L. B. Smith; B. R. Sidner; and T. R. McKee. 1979. Intraslope basin in northwest Gulf of Mexico. *Amer. Assoc. Petrol. Geol., Stud. Geol. No. 7*, p. 289-302.

Brasier, M. D. 1975a. Ecology of Recent sediment-dwelling and phytal foraminifera from the lagoons of Barbuda, West Indies. *Jour. Foram. Res*. 5:42-62.

Brasier, M. D. 1975b. The ecology and distribution of Recent foraminifera from reefs and shoals around Barbuda, West Indies. *Jour. Foram. Res*. 5:193-210.

Bremer, M. L.; A. S. Edwards; and R. C. Thunell. 1979. Benthonic foraminiferal biofacies in silled oceanic basins of the eastern Atlantic: evidence for primary control of their distribution by degree of carbonate saturation. *Geol. Soc. America Abst., Program*, p. 393-394.

Bretsky, P. W., and D. M. Lorenz. 1970. Adaptive response to environmental stability: a unifying concept in paleoecology. *North Amer. Paleont. Convention, Chicago, 1969, Proc., Pt. E*, p. 522-550.

Briggs, J. C. 1974. *Marine zoogeography*. New York: McGraw-Hill, 475 p.

Bright, T. J., and L. H. Pequegnat, eds. 1974. *Biota of West Flower Garden Bank*. Houston: Gulf Publ., 435 p.

Bryant, W. R.; J. W. Antoine; M. Ewing; and B. R. Jones. 1968. Structure of Mexican Continental Shelf and Slope, Gulf of Mexico. *Amer. Assoc. Petrol. Geol., Bull*., 52:1204-1228.

Bryant, W. R.; A. A. Meyerhoff; N. K. Brown; M. A. Furrer; T. E. Pyle; and J. W. Antoine. 1969. Escarpments, reef trends, and diapiric structures, eastern Gulf of Mexico. *Amer. Assoc. Petrol. Geol., Bull*., 53:2506-2542.

Buffler, R. T.; F. J. Shaub; J. S. Watkins; and J. L. Worzel. 1979. Anatomy of the Mexican ridges, southwestern Gulf of Mexico. *Amer. Assoc. Petrol. Geol., Mem. 29*, p. 319-327.

Buzas, M. A. 1967. An application of canonical analysis as a method of comparing faunal areas. *Jour. Animal Ecol*. 36:563-577.

Buzas, M. A., and T. G. Gibson. 1969. Species diversity: benthonic foraminifera in western North Atlantic. *Science* 163:72-75.

Carder, K. L.; K. A. Fanning; P. R. Betzer; and V. Maynard. 1977. Dissolved silica and the circulation in the Yucatan Strait and the deep eastern Gulf of Mexico. *Deep-Sea Res*. 44:1149-1160.

Cushman, J. A. 1922. Shallow-water foraminifera of the Tortugas region. *Carnegie Inst., Washington, Publ. 311 (Dept. Marine Biol., Papers*, v. 17), 85 p.

Davies, D. K. 1972. Deep-sea sediments and their sedimentation, Gulf of Mexico. *Amer. Assoc. Petrol. Geol., Bull*. 56:2212-2239.

Davies, D. K., and W. R. Moore. 1970. Dispersal of Mississippi sediment in the Gulf of Mexico. *Jour. Sed. Petrol*. 40:339-353.

Davis, R. A., Jr. 1964. Foraminiferal assemblages of Alacran Reef, Campeche Bank, Mexico. *Jour. Paleont.* 38:417-421.

Dayton, P. K., and R. R. Hessler. 1972. Role of biological disturbance in maintaining diversity in the deep sea. *Deep-Sea Res.* 19: 199-208.

Dignes, T. W. 1978. Late Quaternary abyssal foraminifera of the Gulf of Mexico. Ph.D. thesis, University of Maine.

Douglas, R. G. 1979. Benthic foraminiferal ecology and paleoecology: a review of concepts and methods. In J. H. Lipps; W. H. Berger; M. A. Buzas; R. G. Douglas; and C. A. Ross. *Foraminiferal ecology and paleoecology.* Soc. Econ. Paleont. Mineral., Short Course No. 6, Houston, Texas, 1979, p. 21-23.

Douglas, R. G.; and H. L. Heitman. 1979. Slope and basin benthic foraminifera of the California borderland. *Soc. Econ. Paleont. Mineral., Spec. Publ. No. 27,* p. 231-246.

Douglas, R. G.; and F. Woodruff. 1980. Deep-sea foraminifera. In C. Emiliani, (ed.), *The oceanic lithosphere: The Sea.* New York: Wiley-Interscience, vol. 7. (in press).

Elton, C. 1927. *Animal ecology.* New York: Macmillan, 207 p.

Ewing, M.; D. B. Ericson; and B. C. Heezen. 1958. Sediments and topography of the Gulf of Mexico. In L. G. Weeks, ed., Habitat of oil - A symposium: *Amer. Assoc. Petrol. Geol.,* Tulsa, p. 955-1053.

Flint, J. M. 1899. Recent foraminifera: A descriptive catalogue of specimens dredged by the U.S. Fish Commission Steamer *Albatross. U.S. Nat. Mus. Ann. Rept. (1897),* pt. 1, p. 249-349.

Garrison, L. E., and R. G. Martin. 1973. Geologic structures in the Gulf of Mexico Basin. *U.S. Geol. Survey, Prof. Paper 773,* 85 p.

Gibson, T. G., and M. A. Buzas. 1973. Species diversity: patterns in modern and Miocene foraminifera of the eastern margin of North America. *Geol. Soc. Amer. Bull.* 84:217-238.

Goës, A. 1896. Reports on the dredging operations off the west coast of Central America to the Galapagos, to the west coast of Mexico, and in the Gulf of California, in charge of Alexander Agassiz carried on by the U.S. Fish Commission Steamer *Albatross* during 1891, Lieutenant Commander Z. L. Tanner, U.S.N., Commanding. XX, The foraminifera. *Mus. Comp. Zool., Harvard, Bull.* 29, 104 p.

Gould, H. R., and R. H. Stewart, Jr. 1955. Continental terrace sediments in the northeastern Gulf of Mexico. In J. L. Hough, and H. W. Menard, eds., *Finding ancient shorelines.* Soc. Econ. Paleont. Mineral., Spec. Publ. No. 3, p. 2-20.

Grassle, J. F., and H. L. Sanders. 1973. Life histories and the role of disturbance. *Deep-Sea Res.* 29:643-659.

Greiner, G. O. G. 1974. Environmental factors controlling the distribution of Recent benthonic foraminifera. *Breviora,* no. 420, p. 1-35.

Hackathorn, M. 1975. Foraminifera and some associated microorganisms of Lemon Bay, Florida. Master's thesis, Bowling Green University.

Hedges, J. I., and D. L. Parker. 1976. Faunal-derived organic matter in surface sediments from the Gulf of Mexico. *Geochim. et Cosmochim. Acta* 40:1019-1029.

Hedgpeth, J. W. 1953. An introduction to the zoogeography of the northwestern Gulf of Mexico with reference to the invertebrate fauna. *Publ. Inst. Mar. Sci., Univ. Texas* 3:107-224.

Howard, J. F. 1965. Shallow-water foraminiferal distribution near Big Pine Key, southern Florida Keys. *Compass* 42:265-280.

Huang, T-C., and H. G. Goodell. 1970. Sediments and sedimentary processes of eastern Mississippi Cone. *Amer. Assoc. Petrol. Geol., Bull.* 54:2070-2100.

Hueni, C.; J. Monroe; J. Gevirtz; and R. Casey. 1978. Distribution of living benthonic foraminifera as indicators of oceanographic processes on the south Texas outer continental shelf. *Gulf Coast Assoc. Geol. Socs., Trans.* 28: 193-200.

Hurley, R. J., and L. K. Fink. 1963. Ripple marks show that countercurrent exists in Florida Straits. *Science* 139:603-605.

Huse, J. M. 1976. Distribution of Recent benthonic foraminifera on the northeast Gulf of Mexico Continental Shelf in relation to character of substrate. Master's thesis, University of South Florida.

Jackson, J. B. C. 1972. The ecology of the molluscs of *Thallasia* communities, Jamaica, West Indies. II. Molluscan population variability along an environmental stress gradient. *Marine Biol.* 14:304-337.

Jackson, J. B. C. 1973. The ecology of the molluscs of *Thallasia* communities, Jamaica, West Indies. I. Distribution, environmental physiology, and ecology of common shallow-water species. *Bull. Marine Sci.* 23:313-350.

Kane, H. E. 1962. The distribution of the foraminifera of Sabine Lake and Sabine Pass, Texas and Louisiana, and the nearshore area of the Gulf of Mexico. *Jour. Paleont.* 36:834-836.

Kane, H. E. 1967. Recent microfaunal biofacies in Sabine Lake and environs, Texas and Louisiana. *Jour. Paleont.* 41:947-964.

Kauffman, E. G. 1977. Evolutionary rates and biostratigraphy. In E. G. Kauffman, and J. E. Hazel, eds., *Concepts and methods of biostratigraphy:* Stroudsburg, Pa.: Dowden, Hutchinson & Ross, p. 109-141.

Kellough, G. R. 1956. Distribution of foraminifera around a submerged hill in the Gulf of Mexico. *Gulf Coast Assoc. Geol. Socs., Trans.* 6:205-216.

Kennett, J. P., and N. J. Shackleton. 1975. Laurentide ice sheet meltwater recorded in Gulf of Mexico deep-sea cores. *Science* 188:147-148.

Kohout, F. A.; H. R. Henry; and J. E. Banks. 1977. Hydrogeology related to geothermal conditions of the Floridan Plateau. *Fla. Dept. Nat. Res., Bur. Geol., Spec. Publ. No. 21,* p. ix-41.

Koons, C. B., and D. Perry. 1976. Geochemical interpretations of *Alaminos* and *Western Shoal* cruise data. In C. E. Pflum, and W. E. Frerichs, *Gulf of Mexico deep-water foraminifera.* Cushman Found. Foram. Res., Spec. Publ. No. 14, Appendix B, p. 46-50.

Kornfeld, M. M. 1931. Recent littoral foraminifera from Texas and Louisiana. *Contr. Dept. Geol., Stanford Univ.* 1:269-476.

Lagoe, M. B. 1977. Recent benthic foraminifera from the central Arctic Ocean. *Jour. Foram. Res.* 7:106-130.

Lamb, G. M. 1972. Distribution of Holocene foraminifera in Mobile Bay and the effect of salinity changes. *Geol. Survey Alabama, Circ. 82,* p. 1-12.

Lankford, R. R. 1959. Distribution and ecology of foraminifera from east Mississippi Delta margin. *Amer. Assoc. Petrol. Geol., Bull.* 43:2065-2099.

Leavesley, A.; M. Bauer; K. McMillen; and R. Casey. 1978. Living shelled microzooplankton (radiolarians, foraminiferans, and pteropods) as indicators of oceanographic processes in water over the outer continental shelf of south Texas. *Gulf Coast Assoc. Geol. Socs., Trans.* 28:229-238.

Le Calvez, Y. 1977. Révision des Foraminifères de la collection d'Orbigny: II--Foraminifères de l'île de Cuba--Tome 1. *Cah. Micropaléont.* v. 1, 128 p.

Lehner, P. 1969. Salt tectonics and Pleistocene stratigraphy on continental slope of northern Gulf of Mexico. *Amer. Assoc. Petrol. Geol., Bull.* 53:2431-2479.

LeRoy, D. O., and S. A. Levinson. 1974. A deep-water Pleistocene microfossil assemblage from a well in the northern Gulf of Mexico. *Micropaleontology* 20:1-37.

LeRoy, D. O., and K. A. Hodgkinson. 1975. Benthonic foraminifera and some pteropoda from a deep-water dredge sample, northern Gulf of Mexico. *Micropaleontology* 21: 420-447.

Levins, R. 1968. *Evolution in changing environments.* Monogr. Population Biol., v. 2. Princeton: Princeton University Press, 120 p.

Levinton, J. S. 1970. The paleoecological significance of opportunistic species. *Lethaia* 3:69-78.

Levinton, J. S. 1973. Genetic variation in a gradient of environmental variability: marine

Bivalvia (Mollusca). *Science* 180:75-76.
Levinton, J. S. 1974. Trophic group and evolution in bivalve molluscs. *Palaeontology* 7:579-585.
Lidz, L., and B. Lidz. 1966. Foraminiferal biofacies of Veracruz reefs. *Amer. Assoc. Petrol. Geol., Bull.* 50:1514-1517.
Loeblich, A. R., Jr., and H. Tappan. 1964. Protista 2, Sarcodina, chiefly "thecamoebians" and Foraminiferida: *Treatise on Invertebrate Paleontology*, Geol. Soc. America and Univ. Kansas Press, Lawrence, pt. C, v. 1, 510 p.; v. 2, 389 p.
Loep, K. J. 1965. A study of ecology and distribution of Recent foraminifera in the northwest Gulf of Mexico. *Gulf Coast Assoc. Geol. Socs., Trans.* 15:167-177.
Logan, B. W., et al. 1969. Carbonate sediments and reefs, Yucatan Shelf, Mexico. *Amer. Assoc. Petrol. Geol., Mem. 11*, 198 p.
Lohman, G. P. 1978. Abyssal benthonic foraminifera as hydrographic indicators in the western South Atlantic Ocean. *Jour. Foram. Res.* 8:6-34.
Lowman, S. W. 1949. Sedimentary facies in Gulf Coast. *Amer. Assoc. Petrol. Geol., Bull.* 33:1939-1997.
Ludwick, J. C., and W. R. Walton. 1957. Shelf-edge calcareous prominences in northeastern Gulf of Mexico. *Amer. Assoc. Petrol. Geol., Bull.* 41:2054-2101.
Lynch, S. A. 1954. Geology of the Gulf of Mexico. *U.S. Fish. Bull.* 89:67-85.
Lynts, G. W. 1962. Distribution of Recent foraminifera in upper Florida Bay and associated sounds. *Cushman Found. Foram. Res., Contr.* 13:127-144.
Lynts, G. W. 1965. Observations on some Recent Florida Bay foraminifera. *Cushman Found. Foram. Res., Contr.* 16:67-69.
Martin, R. G., and A. H. Bouma. 1979. Physiography of Gulf of Mexico. *Amer. Assoc. Petrol. Geol., Stud. Geol.*, no. 7, p. 3-19.
Mello, J. E., and M. A. Buzas. 1968. An application of cluster analysis as a method of determining biofacies. *Jour. Paleont.* 42:747-758.
Moody, C. L. 1967. Gulf of Mexico distributive province. *Amer. Assoc. Petrol. Geol., Bull.* 51: 179-199.
Moore, G. T. 1973. Submarine current measurements - northwest Gulf of Mexico. *Gulf Coast Assoc. Geol. Socs., Trans.* 23:245-255.
Moore, W. E., 1957. Ecology of Recent foraminifera in northern Florida Keys. *Amer. Assoc. Petrol. Geol., Bull.* 41:727-741.
Murray, G. E. 1961. *Geology of the Atlantic and Gulf coastal province of North America.* New York: Harper and Bros., 692 p.
Murray, J. W. 1963. Ecological experiments on foraminiferida. *Jour. Mar. Biol. Assoc. U.K.* 43: 621-642.
Murray, J. W. 1973. *Distribution and ecology of living benthic foraminiferids.* London: Heinemann Educational Books, 274 p.
Nelson, H. F., and E. E. Bray. 1970. Stratigraphy and history of the Holocene sediments in the Sabine-High Island area, Gulf of Mexico. In J. P. Morgan and R. H. Shaver, eds., *Deltaic sedimentation, modern and ancient.* Soc. Econ. Paleont. Mineral., Spec. Publ. 15, p. 48-77.
Neumann, A. C., and M. M. Ball. 1970. Submersible observations in the Straits of Florida: Geology and bottom currents. *Geol. Soc. America Bull.* 81:2861-2874.
Nowlin, W. D., Jr. 1971. Water masses and general circulation of the Gulf of Mexico. *Oceanol. Internat.* (Feb.):28-33.
Osman, R. W., and R. B. Whitlatch. 1978. Patterns of species diversity: Fact or artifact? *Paleobiology* 4:41-54.
Otvos, E. G., Jr. 1978. Calcareous benthic foraminiferal fauna in a very low salinity setting, Lake Pontchartrain, Louisiana. *Jour. Foram. Res.* 8:262-269.
Pachut, J. R., and R. L. Anstey. 1979. A developmental explanation of stability-diversity-variation hypotheses: morphogenetic regulation in Ordovician bryozoan colonies. *Paleobiology* 5:168-187.
Parker, F. L. 1954. Distribution of the foraminifera in the northeastern Gulf of Mexico. *Mus.*

Comp. Zool., Harvard, Bull. 111: 453-588.

Parker, F. L. 1958. Eastern Mediterranean foraminifera. *Swedish Deep-Sea Exped., Rept.* 8:219-283.

Parker, F. L.; F. B. Phleger; and J. F. Peirson. 1953. *Ecology of foraminifera from San Antonio Bay and environs, southwest Texas.* Cushman Found. Foram. Res., Spec. Publ. No. 2, 72 p.

Parker, R. H. 1975. *The study of benthic communities: A model and review.* New York: Elsevier Scientific, 279 p.

Pequegnat, W. E. 1972. A deep bottom current on the Mississippi Cone. In L. R. A. Capurro and J. L. Reid, eds., *Contributions on the physical oceanography of the Gulf of Mexico.* Texas A&M Univ. Oceanogr. Ser. 2:65-87.

Pequegnat, W. E.; B. M. James; A. H. Bouma; W. R. Bryant; and A. D. Fredericks. 1972. Photographic study of deep-sea environments of the Gulf of Mexico. In R. Rezak and V. J. Henry, eds., *Contributions on the geological and geophysical oceanography of the Gulf of Mexico.* Texas A&M Univ. Oceanogr. Ser. 3:67-128.

Pflum, C. E., and W. E. Frerichs. 1976. *Gulf of Mexico Deep-water foraminifers.* Cushman Found. Foram. Res., Spec. Publ. No. 14, 125 p.

Phleger, F. B. 1954. Foraminifera and associated organisms from Mississippi Sound and environs. *Amer. Assoc. Petrol. Geol., Bull.* 38:584-647.

Phleger, F. B. 1955. Ecology of foraminifera in southeastern Mississippi Delta area. *Amer. Assoc. Petrol. Geol., Bull.* 39: 712-752.

Phleger, F. B. 1956. Significance of living foraminiferal populations along the central Texas coast. *Cushman Found. Foram. Res., Contr.* 7:106-151.

Phleger, F. B. 1960a. Sedimentary patterns of microfaunas in northern Gulf of Mexico. In F. P. Shepard et al., eds., *Recent sediments, northwest Gulf of Mexico.* Tulsa: Amer. Assoc. Petrol. Geol., p. 267-301.

Phleger, F. B. 1960b. Foraminiferal populations in Laguna Madre, Texas. *Tohoku Univ. Sci. Repts., 2d Ser. Geology, Spec. Vol. No. 4,* p. 83-91.

Phleger, F. B. 1960c. *Ecology and distribution of Recent foraminifera.* Baltimore: Johns Hopkins Press, 297 p.

Phleger, F. B. and R. R. Lankford. 1957. Seasonal occurrences of living foraminifera in some Texas bays. *Cushman Found. Foram. Res., Contr.,* 8:93-105.

Phleger, F. B. and R. R. Lankford. 1978. Foraminifera and ecological processes in the Alvarado Lagoon area, Mexico. *Jour. Foram. Res.* 8:127-131.

Phleger, F. B. and F. L. Parker, 1951. Ecology of foraminifera, northwest Gulf of Mexico. *Geol. Soc. America, Mem. 46,* pt. 1: 1-88, pt. 2:1-64.

Poag, C. W. 1972. Shelf-edge submarine banks in the Gulf of Mexico: Paleoecology and biostratigraphy. *Gulf Coast Assoc. Geol. Socs., Trans.* 22:267-287.

Poag, C. W. 1973. Late Quaternary sea levels in the Gulf of Mexico. *Gulf Coast Assoc. Geol. Socs., Trans.* 23:394-400.

Poag, C. W. 1976. The foraminiferal community of San Antonio Bay. In A. H. Bouma, ed., *Shell dredging and its influence on Gulf Coast environments.* Houston: Gulf Publ., p. 304-336.

Poag, C. W. 1978. Paired foraminiferal ecophenotypes in Gulf Coast estuaries: Ecological and paleoecological implications. *Gulf Coast Assoc. Geol. Socs., Trans.* 28:pt. II, p. 395-421.

Poag, C. W. 1980. Foraminiferal stratigraphy, paleoenvironments, and depositional cycles in the outer Baltimore Canyon Trough. *U.S. Geol. Survey Circular 833,* p. 44-65.

Poag, C. W., and W. E. Sweet, Jr. 1972. Claypile Bank, Texas Continental Shelf. In R. Rezak, and V. J. Henry, eds., *Contributions on the geological and geophysical oceanography of the Gulf of Mexico.* Texas A&M Univ. Oceanogr. Ser. 3:223-261.

Poag, C. W., and R. C. Tresslar. 1979. Habitat associations among living foraminifers of West Flower Garden Bank, Texas Continental Shelf. *Gulf Coast Assoc. Geol. Socs., Trans.* 29:347-351.

Poag, C. W., and R. C. Tresslar. 1981. Living foraminifers of West Flower Garden Bank, northernmost coral reef in the Gulf of Mexico. *Micropaleontology* (in press).

Poag, C. W., and P. C. Valentine. 1976. Biostratigraphy and ecostratigraphy of the Pleistocene basin, Texas-Louisiana Continental Shelf. *Gulf Coast Assoc. Geol. Socs., Trans.* 26:185-256.

Post, R. J. 1951. Foraminifera of the south Texas coast. *Inst. Mar. Sci., Publ.* 2:165-176.

Pulley, T. E. 1952. A zoogeographic study based on the bivalves of the Gulf of Mexico. Ph.D. thesis, Harvard University.

Pyle, T. E.; J. W. Antoine; D. A. Fahlquist; and W. R. Bryant. 1969. Magnetic anomalies in Straits of Florida. *Amer. Assoc. Petrol. Geol., Bull.* 53:2501-2505.

Robinson, G. S., and B. Kohl. 1978. Computer assisted paleoecologic analyses and application to petroleum exploration. *Gulf Coast Assoc. Geol. Socs., Trans.* 28: pt. II, p. 433-447.

Rowe, G. T., and D. W. Menzel. 1971. Quantitative benthic samples from the deep Gulf of Mexico with some comments on the measurement of the deep-sea biomass. *Bull. Marine Sci.* 21:556-566.

Rusnak, G. A. 1960. Sediments of Laguna Madre, Texas. In F. P. Shepard et al., eds. *Recent sediments, northwest Gulf of Mexico.* Tulsa: Amer. Assoc. Petrol. Geol., p. 153-195.

Sanders, H. L. 1968. Marine benthic diversity: a comparative study. *Amer. Nat.* 102:243-282.

Segura, L. R. 1963. *Sistematica y distribucion de los Foraminifera litorales de "Playa Washington" al sureste de Matamoros, Tamaulipas, Mexico.* Univ. Nal. Auton. Mexico, Inst. Geol., Bol. 68, 92 p.

Sen Gupta, B. K. 1977. Depth-distribution of modern benthic foraminifera on continental shelves of the world ocean: *Indian Jour. Earth Sci.* 4:60-83.

Shepard, F. P.; F. B. Phleger; and Tj. H. van Andel, eds. 1960. *Recent sediments, northwest Gulf of Mexico.* Tulsa: Amer. Assoc. Petrol. Geol., 394 p.

Shifflett, E. 1961. Living, dead, and total foraminiferal faunas, Heald Bank, Gulf of Mexico. *Micropaleontology* 7:45-54.

Sidner, B. R.; S. Gartner; and W. R. Bryant. 1979. Late Pleistocene geologic history of Texas outer continental shelf and upper continental slope. *Amer. Assoc. Petrol. Geol., Stud. Geol. No. 7,* p. 243-266.

Sidner, B. R., and T. T. McKee. 1976. Geochemical controls on vertical distribution of iron-rich agglutinated foraminifers in Late Quaternary Continental Slope sediments from northwest Gulf of Mexico. *Amer. Assoc. Petrol. Geol., Bull.* 60:722.

Sliter, W. V., and R. A. Baker. 1972. Cretaceous bathymetric distribution of benthic foraminifera. *Jour. Foram. Res.* 2:167-183.

Somero, G. N., and M. Soulé. 1974. Genetic variation in marine fishes as a test of the niche-variation hypothesis. *Nature 249:670-672.*

Sorenson, F. H.; L. W. Snodgrass; J. H. Rebman; R. R. Murchison; and C. R. Jones. 1975. Preliminary bathymetric map of Gulf of Mexico region. *U.S. Geol. Survey Open-file Map 75-140,* scale 1:2,500,000.

Stuart, C. J., and C. A. Caughey. 1977. Seismic facies and sedimentology of terrigenous Pleistocene deposits in northwest and central Gulf of Mexico. *Amer. Assoc. Petrol. Geol., Mem. 26,* p. 249-275.

Sturges, W., and J. P. Blaha. 1975. A western boundary current in the Gulf of Mexico. *Science* 192: 367-369.

Thunnell, R. C. 1976. Calcium carbonate dissolution history in Late Quaternary deep-sea sediments, western Gulf of Mexico. *Quat. Res.* 6:281-297.

Trask, P. D. 1953. The sediments of the western Gulf of Mexico, Part II--Chemical studies of the western Gulf of Mexico. *Papers in Phys. Oceanogr., Meteorol.,*

Mass. Inst. Tech., Woods Hole Oceanogr. Inst. 12:45-120.

Tresslar, R. C. 1974. The living benthonic foraminiferal fauna of the West Flower Garden Bank coral reef and biostrome. Master's thesis, Texas A&M University.

Trewartha, G. T. 1954. *An introduction to climate.* New York: McGraw-Hill, 402 p.

Upshaw, C. F.; W. B. Creath; and F. L. Brooks. 1966. *Sediments and microfauna off the coasts of Mississippi and adjacent states.* Mississippi Geol. Econ. Topog. Survey, Bull. 106, 127 p.

van Andel, Tj. H. 1960. Sources and dispersion of Holocene sediments, northern Gulf of Mexico. In F. P. Shepard et al., eds., *Recent sediments, northwest Gulf of Mexico.* Tulsa: Amer. Assoc. Petrol. Geol., p. 267-301.

Waldron, R. P. 1963. *A seasonal ecological study of foraminifera from Timbalier Bay, Louisiana.* Louisiana State Univ. Coastal Stud. Inst., Contr. No. 63-3, 57 p.

Walton, W. R. 1964a. Recent foraminiferal ecology and paleoecology. In J. Imbrie and N. Newell, eds., *Approaches to paleoecology.* New York: John Wiley, p. 151-237.

Walton, W. R. 1964b. Ecology of benthic foraminifera in the Tampa-Sarasota Bay area, Florida. In R. L. Miller, ed., *Papers in marine geology,* Shepard Commemorative Volume. New York: Macmillan, p. 429-454.

Wantland, K. F. 1969. Distribution of modern brackish-water Foraminifera in Trinity Bay. In R. R. Lankford and J. J. W. Rogers, eds., *Holocene geology of the Galveston Bay area.* Houston: Houston Geol. Soc., p. 93-117.

Warren, A. D. 1957. Foraminifera of the Buras-Scofield Bayou region southeast Louisiana. *Cushman Found. Foram. Res., Contr.* 8:29-40.

Watkins, J. S.; J. W. Ladd; R. T. Buffler; F. J. Shaub; M. H. Houston; and J. L. Worzel. 1979. Occurrence and evolution of salt in deep Gulf of Mexico. *Amer. Assoc. Petrol. Geol., Stud. Geol. No. 7,* p. 43-65.

Whittington, H. B. 1964. Taxonomic basis of paleoecology. In J. Imbrie and N. Newell, eds., *Approaches to paleoecology.* New York: John Wiley, p. 19-31.

Plate Index

Plate

3. *A. pauciloculata*, 134, 135
4. *Homotrema rubra*, 134, 135

47-48, Fig. 1. *Asterigerina carinata*, 136, 137
2. *Amphistegina gibbosa*, 136, 137
3. *Peneroplis proteus*, 136, 137
4. *Archaias angulatus*, 136, 137

49-50, Fig. 1. *Saccammina difflugiformis* f. *typica*, 138, 139
2. *Nouria polymorphinoides*, 138, 139
3. *Ammobaculites* sp., 138, 139
4. *A.* sp. cf. *A. foliaceus*, 138, 139

51-52, Fig. 1. *Ammotium salsum* f. *dilatatum*, 140, 141
2. *A. salsum* f. *fragile*, 140, 141
3. *A. salsum* f. *typicum*, 140, 141
4. *A. salsum* f. *exile*, 140, 141

53-54, Fig. 1. *Spiroloculina soldanii* f. *dentata*, 142, 143
2. *S. soldanii* f. *typica*, 142, 143
3. *S. antillarum* f. *crenulata*, 142, 143
4. *Wiesnerella auriculata*, 142, 143

55-56, Fig. 1. *Quinqueloculina compta*, 144, 145
2. *Q. parkeri* f. *occidentalis*, 144, 145
3. *Q. bicarinata*, 144, 145
4. *Q. candeiana*, 144, 145

Plate

57-58, Fig. 1. *Triloculina variolata*, 146, 147
2. *T. trigonula*, 146, 147
3. *T. tricarinata*, 146, 147
4. *T.* sp., 146, 147

59-60, Fig. 1. *Quinqueloculina agglutinans*, 148, 149
2. *Q. horrida*, 148, 149
3. *Miliolinella circularis*, 148, 149
4. *M. fichteliana*, 148, 149

61-62, Fig. 1. *Quinqueloculina crassa* f. *subcuneata*, 150, 151
2. *Q. linneiana* f. *comis*, 150, 151
3. *Articulina pacifica*, 150, 151
4. *Nodobaculariella cassis*, 150, 151

63-64, Fig. 1. *Quinqueloculina bicostata* f. *typica*, 152, 153
2. *Q. bicostata* f. *garrettii*, 152. 153
3. *Q. collumnosa* f. *excavata*, 152, 153
4. *Q. tipswordi*, 152, 153

Subject Index

About the Author

C. WYLIE POAG is a micropaleontologist specializing in the ecology, paleoecology, and biostratigraphy of benthic and planktic foraminifers. He received the B.S. in Geology from Florida State University in 1959, the M.S. in Geology from Louisiana State University in 1963, and the Ph.D. in Paleontology from Tulane University in 1971.

Dr. Poag joined Chevron Oil Company in 1963 and served as an exploration micropaleontologist in southwest and offshore Louisiana until 1970. For two months in 1970 he was aboard DV *Glomar Challenger* as a foraminiferal specialist during Leg XI of NSF's Deep Sea Drilling Project. He joined the faculty of the Department of Oceanography at Texas A&M University as an assistant professor in 1969, where he taught graduate courses in micropaleontology, marine biostratigraphy, and the history of ocean basins. His research stressed the paleoecology and biostratigraphy of foraminifers in shelf sediments and deep-sea cores from the Gulf of Mexico. He spent 1974 on assignment from Texas A&M to the National Science Foundation as assistant project officer for the Deep Sea Drilling Project. He joined the U.S. Geological Survey in 1974 as a member of the Paleontology and Stratigraphy Branch assigned to the Atlantic and Gulf of Mexico Branch of the Office of Marine Geology in Woods Hole, Massachusetts. In 1976, he served as foraminiferal specialist and paleontological coordinator during the USGS's Atlantic Margin Coring Project. In 1981, he served as co-chief scientist aboard DV *Glomar Challenger* for Leg 80 of the International Phase of Ocean Drilling (IPOD).

Dr. Poag's current research includes analysis of Cenozoic and Mesozoic foraminiferal assemblages and reconstruction of the depositional sequences that built the Atlantic and Gulf of Mexico continental shelf and slope. In addition to his own studies, he is responsible for coordination of other paleontologic and stratigraphic studies of the Atlantic and Gulf coastal plain, shelf, and slope.

He has published numerous technical papers and abstracts and is a member of the American Association of Petroleum Geologists, American Association for the Advancement of Science, Society of Economic Paleontologists and Mineralogists, Paleontology Society, American Quaternary Association, Paleontological Research Institution, and Sigma Xi. He was president of the Gulf Coast Section of the Society of Economic Paleontologists and Mineralogists during 1972-1973.

Errata

Pages 163, 35, 164, 84, 157, 161, 160

Fig. 19

Pl. 30